THE BILINGUAL SERIES OF
THE MOST IMPRESSIVE BEAUTY OF CHINA

最美中国双语系列

科技成就

ACHIEVEMENTS IN SCIENCE AND TECHNOLOGY

主　编◎青　田
副主编◎冉玉体　胡　鹏
参　编◎吉　慧　王海云　王胜利

中国科学技术大学出版社

内容简介

"最美中国双语系列"是一套精品文化推广图书,包括《风景名胜》《民俗文化》《饮食文化》《杰出人物》《科技成就》《中国故事》六册,旨在传播中华优秀文化,传承中华民族宝贵的民族精神,展示奋进中的最美中国,可供广大中华文化爱好者、英语学习者及外国友人参考使用。

本书介绍了中国古代与现代的部分代表性科技成就,展示了中国在科技方面取得的丰硕成果。

图书在版编目(CIP)数据

科技成就:英汉对照/青闰主编.—合肥:中国科学技术大学出版社,2021.11

(最美中国双语系列)

ISBN 978-7-312-05209-5

Ⅰ.科… Ⅱ.青… Ⅲ.科学技术—技术发展—成就—中国—英、汉 Ⅳ.N12

中国版本图书馆CIP数据核字(2021)第121011号

科技成就
KEJI CHENGJIU

出版	中国科学技术大学出版社 安徽省合肥市金寨路96号,230026 http://press.ustc.edu.cn https://zgkxjsdxcbs.tmall.com
印刷	合肥市宏基印刷有限公司
发行	中国科学技术大学出版社
经销	全国新华书店
开本	880 mm×1230 mm 1/32
印张	6.375
字数	153千
版次	2021年11月第1版
印次	2021年11月第1次印刷
定价	35.00元

前　言　Preface

　　文化是一个国家与民族的灵魂。"最美中国双语系列"旨在弘扬和推广中华优秀文化，突出文化鲜活主题，彰显文化核心理念，挖掘文化内在元素，拓展文化宽广视野，为广大读者了解、体验和传播中华文化精髓提供全新的视角。本系列图书秉持全面、凝练、准确、实用、自然、流畅的撰写原则，全方位、多层面、多角度地展现中华文化的源远流长和博大精深，对于全民文化素质的提升具有独特的现实意义，同时也为世界文化的互联互通提供必要的借鉴和可靠的参考。

　　"最美中国双语系列"包括《风景名胜》《民俗文化》《饮食文化》《杰出人物》《科技成就》《中国故事》六册，每册中的各篇文章以文化剪影为主线，以佳句点睛、情景对话和生词注解为副线，别出心裁，精彩呈现中华文化的方方面面。

　　"最美中国双语系列"充分体现以读者为中心的编写理念，从文化剪影到生词注解，读者可由简及繁、由繁及精、由精及思地感知中华文化的独特魅力。书中的主线和副线是一体两面的有机结合，不可分割，如果说主线是灵魂，副线则是灵魂的眼睛。

　　"最美中国双语系列"的推出，是讲好中国故事、展现中国立场、传播中国文化的一道盛宴，读者可以从中感悟生活。

　　《科技成就》包括古代科技成就与现代科技成就两大部分。在中国这片辽阔的土地上，有天文历法、地动仪、四大发明、圆周率、《九章

科技成就

算术》、《天工开物》、中医学、赵州桥、都江堰,有《墨经》、《营造法式》、《考工记》、《梦溪笔谈》、《徐霞客游记》,也有超级杂交水稻、人工合成牛胰岛素、"两弹一星"、汉字激光照排技术、中国航天工程、南京长江大桥、中国高铁、三峡工程,还有"中国天眼"、超级计算机、北斗卫星导航系统、"墨子号"量子通信卫星、"悟空号"暗物质探测卫星、国产大飞机C919……可谓上天入地,纵贯古今,成就斐然,熠熠生辉。

本书由河南理工大学冉玉体、吉慧撰写初稿,焦作大学胡鹏、王海云撰写二稿,上海工程大学王胜利撰写三稿,焦作大学青闰负责全书统稿与定稿。

最后,在本书即将付梓之际,衷心感谢中国科学技术大学出版社的大力支持,感谢朋友们的一路陪伴,感谢家人们始终不渝的鼓励和支持。

青 闰

2021年3月6日

目 录 Contents

前言 Preface ··· i

第一部分 古代科技成就
Part I Ancient Achievements in Science and Technology

天文历法　The Astronomical Calendar ································003
张衡与地动仪　Zhang Heng and Seismometer ····················009
四大发明　The Four Great Inventions ································015
祖冲之与圆周率　Zu Chongzhi and π ·································022
《九章算术》　The Nine Chapters on the Art of Mathematics ········028
《齐民要术》　The Important Arts for the People's Welfare ········034
《天工开物》　An Encyclopedia of Technology ··················041
赵州桥　Zhaozhou Bridge ··047
都江堰　Dujiangyan ···053
《墨经》　The Mohist Canon ···059
《营造法式》　The Rules of Architecture ··························064
《考工记》　The Book of Diverse Crafts ····························069
《梦溪笔谈》　Brush Talks from Dream Brook ···················075
《徐霞客游记》　Xu Xiake's Travels ·································080

第二部分 现代科技成就
Part II Modern Achievements in Science and Technology

超级杂交水稻 Super Hybrid Rice ……………………………089
人工合成结晶牛胰岛素 Synthetic Crystalline Bovine Insulin ……095
"两弹一星" "Two Bombs and One Satellite" ………………101
汉字激光照排技术 The Laser Phototypesetting System of
  Chinese Characters ……………………………………107
中国航天工程 China Aerospace Industry ……………………114
南京长江大桥 Nanjing Yangtze River Bridge …………………121
港珠澳大桥 Hong Kong-Zhuhai-Macao Bridge………………127
中国高铁 China High-speed Railway…………………………132
南极科考 Antarctic Expedition …………………………………137
"蛟龙号"潜水器 "Jiaolong" Submersible …………………143
"中国天眼" "China's Eye of Heaven" ………………………148
"天河二号"超级计算机 "Tianhe-2" Supercomputer …………154
"神威·太湖之光"超级计算机 "Sunway TaihuLight"
  Supercomputer ……………………………………………159
北斗卫星导航系统 Beidou Navigation Satellite System …………164
"墨子号"量子通信卫星 "Mozi" Quantum Communication
  Satellite ……………………………………………………170

目 录

"悟空号"暗物质粒子探测卫星 "Wukong" Dark Matter

    Particle Explorer …………………………………………175

国产大飞机C919　Domestically-made Airplane C919 ……………180

5G领跑世界　5G Leading the World ………………………………185

移动支付　Mobile Payment …………………………………………190

第一部分 古代科技成就

Part I Ancient Achievements in Science and Technology

天文历法

The Astronomical Calendar

导入语 Lead-in

中国古代历法采用阴阳合历，即以太阳运动周期为年，以月亮圆缺周期为月，以闰月协调年与月的关系。天文历法是中国古人根据天体运行变化确定季节时令的科学，被广泛应用于日常生活和农业生产中，一直沿用至今。中国传统天文历法离不开阴阳五行学说。五千多年前，中国就已经有了阴阳历。自商代以来，历代都有专门官员负责掌管天文历法。南北朝以后，天文历法的发展出现了官方和民间两种渠道；隋唐时期历法日趋完臻。掌握天文历法知识，使中国古人进一步增强了改造自然的信心。

科技成就

 文化剪影 **Cultural Outline**

The Chinese ancient calendar applied the lunisolar calendar, which **determined**① the twenty-four solar terms according to the changes of the sun's position and the order of the evolution of the surface climate caused by the changes. A year is divided into twelve months by the moon's waxing and waning. In this way, we can not only pay attention to the more **intuitive**② and practical moon cycle, but also guide farming according to the changes of winter and summer in the solar annual cycle.

中国古代历法采用阴阳合历,根据太阳的位置变化和由此产生的地表气候的演变顺序确立了二十四节气,依据月亮的圆缺将一年分为十二个月。这样既可以关注到较为直观、实用的月亮周期,又能依照太阳年度周期的寒暑变化来指导农事。

The Chinese ancient calendar is **meticulous**③ and complex, the Chinese nation is **impartial**④, does not go to extremes and seeks the national character of the golden mean. We can find the source from the lunisolar calendar. The concept of Yin and Yang has become one of the cultural gene of the Chinese nation.

中国古代历法细致复杂,中华民族不偏不倚、不走极端、保持中庸的民族性格,可以从阴阳合历中找到源头。阴阳观念已经成为中华民族的文化基因之一。

The astronomical calendar explains the birth of the universe with the theory of Yin and Yang, and the motion of the **celestial**⑤ bodies with the theory of five elements, mirroring the simple materialism. The ideas of Taiji, Yin and Yang, dialectics, changes and cycles have been the core of traditional Chinese beliefs in heaven, which have profoundly influenced the development and changes of the Chinese society.

天文历法以阴阳学说解释宇宙的诞生,以五行学说解释天体运行,反映了朴素的唯物主义思想。对时空宇宙的基础认知衍生出的太极、阴阳、辩证、变化、循环等观念,已经成为中国传统天道信仰的核心,深刻地影响了中国社会的发展和变迁。

佳句点睛 Punchlines

1. The traditional Chinese calendar is a type of lunisolar calendar, which refers to the calendar that **takes into account**⑥ the relationship between the sun, the moon and the earth.

中国传统历法采用阴阳合历,它是一种兼顾太阳、月亮与地球关系的历法。

2. The Chinese ancients employed the astronomical calendar to guide their daily life and farming.

中国古人运用天文历法来指导日常生活和农业生产。

3. The calendar is one of the most important symbols that human

beings evolved from ignorance to civilization.

历法是人类从蒙昧进入文明的重要标志之一。

情景对话 Situational Dialogue

A: I heard that you're so accomplished in sundials.

B: Far from being accomplished, I just know something about it.

A: Can you tell me something about it?

B: Of course. Sundial is an instrument for measuring time by using the position of the sun. It consists of a pointer projecting the shadow of the sun, a projection surface (i.e. the dial surface) bearing the projection of the index and a scale line on the dial surface.

A: How many kinds of sundials, then?

B: The most common type of sundial is the garden sundial, which allows the shadow to be projected on a time-marked plane. When the sun moves, the time indicated by the shadow changes accordingly. In fact, sundials can be designed on the surface of any object, so that fixed pointers can produce shadows to measure time. Therefore, sundials have many forms, such as horizontal sundials, **equatorial**① sundials and **meridian**② sundials.

A: And what else?

B: Sundial is not only a timing instrument, but also allows people to create historical and cultural associations. It can also bring closer the relationship between man and nature. At the same time, it can convey a strong humanistic atmosphere to people.

A: Thank you. I didn't expect the sundial to be so interesting. I'll do some further research later.

B: OK, my pleasure. Maybe we can set up a sundial interest group someday.

A: 听说你对日晷很有研究？

B: 根本谈不上什么研究，仅仅是了解一些。

A: 你能给我讲讲吗？

B: 当然可以。日晷是利用太阳的位置测量时间的一种仪器，由一根投射太阳阴影的指针、承受指标投影的投影面（即晷面）和晷面上的刻度线组成。

A: 日晷有多少种？

B: 最常见的是庭院日晷，就是让日影投射在一个标有时刻的平面上，当太阳移动时，影子指示的时间也跟着变动。其实，日晷可以设计在任何物体的表面上，通过固定的指针产生阴影来测量时间，所以日晷有多种形式，比如地平式日晷、赤道式日晷和子午式日晷等。

A: 还有呢？

B: 日晷不仅是计时仪器，还能让人们产生历史和文化的联想，也会拉近人与自然的关系，同时可以向人们传达强烈的人文气息。

A: 谢谢你啊。没想到日晷这么有趣。回头我要好好研究研究。

B: 好的，不客气。说不定到时候我们还能建立一个日晷兴趣小组呢。

科技成就

 生词注解 Notes

① determine /dɪˈtɜːmɪn/ vt. 决定;确定

② intuitive /ɪnˈtjuːɪtɪv/ adj. 凭直觉得到的;直觉的

③ meticulous /məˈtɪkjələs/ adj. 一丝不苟的;注意细节的

④ impartial /ɪmˈpɑːʃl/ adj. 公正的;不偏不倚的

⑤ celestial /səˈlestiəl/ adj. 天空的;天上的

⑥ take into account 考虑到;把……计算在内

⑦ equatorial /ˌekwəˈtɔːriəl/ adj. 赤道的;赤道附近的

⑧ meridian /məˈrɪdiən/ n. 子午线;经线

张衡与地动仪

Zhang Heng and Seismometer

 导入语 Lead-in

张衡是东汉著名科学家，他博学多才，精通阴阳、儒学、数学、天文、地理和机械制造等，因发明地动仪而被称为"地震学的鼻祖"。东汉时期，地震频发，人们认为地震是苍天对皇帝失职的惩戒和警示，因此世人需要一种能知晓、明示天地灾异的仪器。张衡认为，地震是阴阳失调引起的，通过调整阴阳，可以把灾害造成的损失降到最低。于是，他开始研究地动仪并于132年研制出世界上第一台测量地震方位的地动仪，这标志着人类摸索出了一条能在运动系统当中测量系统自身运动的途径，即通过悬挂物的惯性来揭示运动系统间的差异。地动仪的发明，进一步加深了人类对自然规律的认识。

科技成就

文化剪影　Cultural Outline

Based on his own experience of the ground shaking and the **swaying**① of **suspended**② objects when earthquake occurred, Zhang Heng realized that the earthquake could be located by **azimuth**③. So, on the principle of how **seismic**④ wave travels and the **inertial**⑤ principle, he invented the first seismometer in the world to determine the direction of the earthquake in 132 AD. At that time, this instrument was used to successfully predict an earthquake that occurred in the western region of China.

张衡基于自己对地震时地面晃动和悬挂物摇摆的感受，认识到可按方位判断地震的位置，他利用地震波的传播和力学的惯性原理，于132年发明了世界上第一台测定地震方位的地动仪。当时，他利用这台仪器成功测报了中国西部地区发生的一次地震。

The **seismometer**⑥ invented by Zhang Heng has eight azimuths. On each azimuth, there is a dragon head with a ball in its mouth and a toad underneath **corresponding**⑦ to each dragon head. When the earthquake occurred in any azimuth direction, the ball of the dragon's mouth in that direction would fall into the toad. In this way, people could know the direction of the earthquake.

张衡发明的地动仪有八个方位，每个方位上均有口含龙珠的龙头，在每个龙头的下方均有一只蟾蜍与其对应，任何一方如发生地震,该方位龙口所含龙珠就会落入蟾蜍的口中。这样,人们就能知道地震发生的方位。

The seismometer invented by Zhang Heng imitated the response of the suspended object to the earthquake on the principle of seismic measurement, embodied the scientific thought of seismic measurement with the suspended object, and borrowed the trigger mechanism of the latch type in the technical thought. The science and technology of this kind of trigger device even gives the modern scientific instruments a lot of meaningful **enlightenment**®.

张衡发明的地动仪在测震原理上模拟了悬挂物对地震的反应,体现了用悬挂物测震的科学思想,在技术思想上借鉴了门闩类的触发机构,这种触发装置的科学技术原理甚至给当今时代的科学仪器许多有意义的启迪。

佳句点睛 Punchlines

1. Zhang Heng was honored as "The Father of Seismology" by inventing the first seismometer in the world.

张衡因发明世界上第一台地动仪而被尊称为"地震学的鼻祖"。

2. The invention of the seismometer deepened human's understanding of the laws of nature.

地动仪的发明加深了人类对自然规律的认识。

3. The seismometer invented by Zhang Heng successfully predicted the earthquake that occurred in Longxi, a place more than one thousand li away from Luoyang.

张衡发明的地动仪成功测报了当时距离洛阳一千多里的陇西发生的地震。

 情景对话 Situational Dialogue

A: Good morning, boys and girls. Welcome to Henan Museum.

B: Good morning, auntie.

A: What you see now is the recovery model of seismometer invented by Zhang Heng. Do you know which dynasty Zhang Heng lived in?

B: He was a great astronomer, mathematician and inventor in the Eastern Han Dynasty.

A: Great. Do you know what he invented?

C: Yes, I do. They were the seismometer, the armillary sphere and the south-pointing cart.

A: Yes. Zhang Heng invented many ingenious devices. The most famous one was the seismometer. So, he was honored as "The Father of Seismology". And now the seismoscope in front of you was made in 2004. Based on the principle of good-working and reasonable structure, Henan Museum and China Earthquake Network Center cooperated to turn the model into an instrument in order to restore its original one.

D: Auntie, where is the seismometer invented by Zhang Heng?

A: Zhang Heng invented the first seismometer in the world in 132 AD. And it was put on the Spirit Terrace (Ling Tai) in the south of Luoyang, the capital of the Han Dynasty, successfully predicting the earthquake occurred in Longxi, a place more than one thousand li away from Luoyang. Boys and girls, let's take a close look at the restored **seismoscope**①.

B: Bravo!

A: 同学们,早上好,欢迎来到河南博物院。

B: 阿姨好。

A: 你们现在看到的是张衡地动仪的复原模型。你们知道张衡是哪个朝代的人吗?

B: 张衡是东汉时期伟大的天文学家、数学家和发明家。

A: 很好,你们知道他有哪些发明吗?

C: 是的,我知道。他的发明成果有地动仪、浑天仪和指南车。

A: 是的。张衡发明了很多精巧的器物,其中最著名的就是地动仪,他也因此被称为"地震学的鼻祖"。目前你们看到的这个地动仪是2004年复原的,河南博物院和中国地震台网中心依据原理正确、结构合理、还原历史的原则,合作将模型制成了仪器。

D: 阿姨,张衡发明的地动仪在哪里?

A: 张衡于132年发明了世界上第一个地动仪,并将其放置在洛阳城南的灵台上,它曾经成功测报了距离洛阳一千多里的陇西发生的地震。同学们，下面让我们仔细看一看复原的地震仪。

B: 太好了！

生词注解　Notes

① sway /sweɪ/　*vt.* 摇动；说服

② suspend /səˈspend/　*vt.* 悬；挂

③ azimuth /ˈæzɪməθ/　*n.* 方位；地平经度

④ seismic /ˈsaɪzmɪk/　*adj.* 地震的；地震引起的

⑤ inertial /ɪˈnɜːʃl/　*adj.* 惯性的

⑥ seismometer /saɪzˈmɒmɪtə/　*n.* 地震仪；地震检波器

⑦ correspond /ˌkɒrəˈspɒnd/　*vi.* 相对应；符合

⑧ enlightenment /ɪnˈlaɪtnmənt/　*n.* 启迪；启发

⑨ seismoscope /ˈsaɪzməskəʊp/　*n.* 验震器；地震波示波仪

四大发明

The Four Great Inventions

 导入语 Lead-in

　　西汉初年,中国发明了造纸术,东汉时期的蔡伦改进了造纸术并制作出"蔡侯纸",造纸术是书写材料的一次伟大革命。隋唐时出现了雕版印刷,868年印制的《金刚经》是世界上现存最早的雕版印刷品,11世纪初北宋毕昇发明的活字印刷术,比欧洲早了四百多年。战

国时,出现了指示方向的仪器"司南",后来人们利用磁石指南原理制成指南针,北宋时期指南针运用于航海,13世纪传入阿拉伯和欧洲。火药由古代炼丹家发明于隋唐时期,唐末运用于军事,南宋时发明突火枪,13世纪传入阿拉伯和欧洲。造纸术、印刷术、指南针和火药的发明,是中国古代人民在劳动过程中创造性的智慧体现,为世界留下了一串光耀千秋的辉煌足迹。在14世纪之前,四大发明以各种途径传播到了欧洲,推动了欧洲社会在文化、思想、航海和政治上的巨大进步,从而促进了整个世界文明的发展。

文化剪影 Cultural Outline

The Four Great Inventions refer to papermaking, printing, compass and gunpowder, which were invented and created by ancient Chinese people in their labor work and in practice. In the process of **communication**[①], the Four Great Inventions have played an important role in promoting the progress of world civilization. They are also greatest inventions that have shaken the world for nearly two thousand years.

四大发明是指造纸术、印刷术、指南针和火药,是中国古代人民在不断劳动和实践过程中创造出来的。在持续的传播和交流过程中,四大发明对世界文明和科学的进步产生了重要的推动作用,也是近两千年来震撼世界的伟大发明之一。

On the basis of silk-making by his **predecessors**[②], Cai Lun of the Eastern Han Dynasty used the fiber of plants to make plant fiber paper suitable for writing, and developed an independent paper making technique. Printing was developed from the seals and stone **engraving**[3], and wood-blocked printing in the Sui and Tang Dynasties. In the Song Dynasty, Bi Sheng improved the traditional types and developed the movable type of printing, making printing faster and easier than before. Compass was developed from the ancient Chinese people's observation on the **magnetic**[4] force of magnetite. The Chinese people made a spoon-like compass in the Han Dynasty. The people further improved it and developed the compass by using the method of artificial magnetization on iron needles in the Northern Song Dynasty. Gunpowder originated from the **alchemy**[5] of Taoism in ancient China. In the Tang Dynasty, people began to use gunpowder for military purposes.

在前人制造丝织的基础上，东汉的蔡伦利用植物纤维制造出适合书写的植物纤维纸，发展了独立的造纸工艺。印刷术始于隋唐时期的雕版印刷，宋朝毕昇对其进了发展和完善，从而产生了活字印刷术。指南针始于中国古代先民对磁现象的观察和研究，汉朝时期，人们制造出了形如勺子的指南针。北宋时期，人们用人工磁化铁针的方法制成了指南针。火药源于中国古代道家的炼丹术，唐朝开始将其用于军事。

The invention and communication of the Four Great Inventions have promoted the progress of world civilization. Papermaking and printing have promoted the publishing industry, culture communication, and civilian education. Gunpowder has developed European **firearms**⑥ and the methods of warfare, promoted the changes in modern European society, and driven the course of human history. The compass has promoted the development of modern **navigation**⑦ in the world, the large-scale economic and cultural exchanges between countries, and the rapid advance of modern world civilization.

造纸术、印刷术、指南针和火药的发明和传播,推动了世界文明的进步。造纸术和印刷术促进了出版业的发展和文化的传播,使平民教育成为可能。指南针推动了世界近代航海事业的发展,促进了各国之间大规模的经济文化交流和世界近代文明的突飞猛进。火药促进了欧洲火器的发展和作战方法的变革,促进了欧洲近代社会的变革,推动了整个人类历史的进程。

佳句点睛 Punchlines

1. The Four Great Inventions refer to China's papermaking, printing, compass and gunpowder.

四大发明是指中国的造纸术、印刷术、指南针和火药。

2. The Four Great Inventions demonstrate the creative wisdom of ancient Chinese in their labor work.

四大发明是中国古代人民在劳动过程中创造性的智慧体现。

3. The communication of the Four Great Inventions has promoted

the progress of world civilization.

四大发明的传播推动了世界文明的进步。

情景对话 Situational Dialogue

A: How about you? We've been walking in the mountains for a whole day.

B: Oops, I'm **exhausted**. And now I have no sense of direction. Around us are the dense forests. I'm wondering whether we are lost.

A: No problem. We can find our way out. I have a piece of magic **gadget**.

B: What's it?

A: Compass.

B: Compass? That's the invention of our ancestors.

A: Yes. Ancient Chinese invented the compass based on their understanding of magnetic force. The earliest compass was known as Si'nan, a lodestone compass, which the ancient Chinese used to search minerals. It's recorded in *Guan Zi* that there are magnets on the mountain, and gold and copper under it. And it's said that the First Emperor of Qin had the Epang Palace built near Xianyang after he united the six States. In the Palace, one gate was made of magnet, which could catch those who carried the weapons to **assassinate** in the palace.

B: We Chinese are actually wise.

A: Of course. We have the Ancient Four Great Inventions as papermaking, printing, compass and gunpowder. And now we have some new inventions as high-speed trains, E-payment, shared bikes and online shopping.

B: Look! That's the way down to the mountain. We've walked out of the mountain. Your gadget does really work.

A: 你还好吗？我们都在山里走了整整一天了。

B: 哎呀,我快累死了,现在我一点方向感都没有了,周围都是茂密的树林,我们是不是迷路了？

A: 没事,我们可以走出大山,我有法宝。

B: 什么法宝？

A: 指南针。

B: 指南针？那是我们老祖先的发明。

A: 是的。指南针的发明基于中国古代劳动人民对物体磁性的认识,最早的指南针是司南,人们用它来探寻矿石。《管子》里记载:"山上有磁石者,其下有金铜。"据说,秦始皇统一六国后,在咸阳附近修建阿房宫,宫中有一座门是用磁石做成的,可以吸住持兵器入宫行刺的人。

B: 看来我们中国人的确很聪明。

A: 当然,我们不仅古代有造纸术、印刷术、指南针和火药这结四大发明,现在我们还有高铁、扫码支付、共享单车和网购等新发明呢。

B: 哎呀！那不是下山的路嘛,我们走出大山了,你的法宝真管用呀！

生词注解 Notes

① communication /kəˌmjuːnɪˈkeɪʃn/ *n.* 沟通；交流

② predecessor /ˈpriːdəsesə(r)/ *n.* 前任；前辈

③ engraving /ɪnˈgreɪvɪŋ/ *n.* 雕刻；雕刻品

④ magnetic /mægˈnetɪk/ *adj.* 磁的；磁性的

⑤ alchemy /ˈælkəmɪ/ *n.* 炼丹术；炼金术

⑥ firearm /ˈfaɪərɑːm/ *n.* 火器；枪炮

⑦ navigation /ˌnævɪˈgeɪʃn/ *n.* 导航；领航

⑧ exhausted /ɪgˈzɔːstɪd/ *adj.* 筋疲力尽的；疲惫不堪的

⑨ gadget /ˈgædʒɪt/ *n.* 小器具；小装置

⑩ assassinate /əˈsæsɪneɪt/ *vt.* (尤为政治目的)暗杀；行刺

祖冲之与圆周率

Zu Chongzhi and π

导入语 Lead-in

祖冲之是中国南北朝时期伟大的科学家,毕生探究自然科学,在数学、天文历法和机械制造方面做出了杰出贡献。他将圆周率计算到小数点后7位,即介于3.1415926和3.1415927之间,是当时世界上最精确的圆周率数值,他提出的"祖率"对数学的研究有重大贡献。他改革立法,制定了《大明历》,明确提出了改闰法和定岁差的内容,是我国历法上的一次重大改革。他成功复原了指南车,研制出了千里船、欹器等先进实用的工具,有力推动了当时科技与社会的进一步发展。

文化剪影　Cultural Outline

Zu Chongzhi, a noted mathematician, astronomer, and machine maker during the Southern and Northern Dynasties, was one of the **iconic**① figures of leading scientists in ancient China. Meanwhile, he was **versed**② in literature and music, regarded as a model of **combination**③ of literature and science in the history of ancient Chinese science.

祖冲之是中国南北朝时期著名的数学家、天文学家和机械制造家，是中国古代科学领先世界的标志性人物之一。同时，他还通晓文学和音乐，堪称中国古代科学史上文理兼备的典范。

On the basis of previous experience and achievements, Zu Chongzhi figured out π value to 3.1415926 and 3.1415927 through the hard work and repeated calculations. Furthermore, he worked out the **approximation**④ rate of π **fraction**⑤ 22/7 and the density 355/113. Such an accurate π value was more than a thousand years earlier than that in the Western mathematical community.

在前人已有经验与成就的基础上，祖冲之大胆创新、刻苦钻研，通过大量、反复的实践与演算，将π值精确到3.1415926与3.1415927之间，又进一步得出了π分数形式的约率22/7以及密率355/113，得出如此精确的π值比西方数学界早了一千多年。

Zu Chongzhi's brilliant achievements in calculating π value reflected the high development of mathematics in China at that time. The **cyclotomy**① adopted by Zu Chongzhi in his calculating involves limit thoughts. He had to calculate a regular **polygon**② in a circle and a large number of complex calculations including the square root on nine digits. This required hard work and superb mathematical skills.

祖冲之在圆周率计算上的辉煌成就反映了当时中国数学的高度发展。祖冲之计算所采纳的割圆术是一种包含极限思想的方法,要算到圆内接正边形,要对9位大的数字进行包括开方在内的大量复杂计算。这需要辛勤的努力和高超的数学技巧。

佳句点睛　Punchlines

1. Zu Chongzhi was a noted mathematician, astronomer, and machine maker during the Southern and Northern Dynasties in China.

祖冲之是中国南北朝时期著名的数学家、天文学家和机械制造家。

2. Zu Chongzhi's brilliant achievements in mathematics fully demonstrated the high level of development of ancient Chinese science.

祖冲之在数学方面的辉煌成就充分表明了中国古代科学的高度发展水平。

3. In honor of his contribution to the precise calculation of π, Milv (density rate) was called Zulv (approximate ratio of circumference of a

circle to its diameter).

国际数学界为纪念祖冲之对圆周率精确计算的贡献,将密率称为"祖率"。

情景对话　Situational Dialogue

A: Daddy, our math teacher asks us to bring a piece of string, a ruler and three round objects to school tomorrow.

B: Oh. It seems that you will learn the **circumference**® of the circle tomorrow?

A: Yes. Our teacher has texted the message to your phone. He also required us to tell the story of Zu Chongzhi and π.

B: Let me check. Oh, yes. First, let's surf the Internet and learn it together.

A: OK.

B: Zu Chongzhi was a wellknown mathematician, astronomer, and machine maker in the Northern and Southern Dynasties, with many outstanding contributions. One of them was the calculation of π. In order to calculate the π value, he drew a large circle with a diameter of 3.333 meters on the floor of his study. In the circle, he made 12,288 polygons from one **hexagon**® and calculated the circumference of these polygons one by one. At that time, the mathematical calcula-tions were counted with bamboo chips, not the Arabic numbers. So it's a very hard work. He worked day and night year by year. Eventually, he figured out that the π value was between 3.1415926 and 3.1415927, with the 7th number after the decimal point. This π value represents the highest level in mathematics in the world at that time. It was until more than one thousand years later that the German mathematician got the same result.

A: Daddy, shall we go to your study to draw a circle with a diameter of 3.333 meters to calculate the π value?

B: I'm afraid that my study is not so big. Shall we cut a small circle?

A: OK, let's start.

A: 爸爸,数学老师要我们明天带一段绳子、一把尺子和三个不同的圆形物体到学校。

B: 是吗?看来明天你们要开始学习圆的周长了。

A: 对啊,老师不是把作业发你手机上了吗?还让我们明天上课讲祖冲之与圆周率的故事呢。

B: 我看看。噢,是的。那我们先上网查查资料,一起学习。

A: 好的。

B: 祖冲之是中国南北朝时期著名的数学家、天文学家机械制造,他有许多卓越的贡献,其中之一就是对圆周率的计算。为了计算圆周率,他在书房的地面上画了一个直径一丈的大圆,从这个圆的内接正六边形一直做到12288边形,然后一个一个算出这些多边形的周长。那时候的数学计算不是用现在的阿拉伯数字,而是用竹片做的筹码计算。他夜以继日、成年累月,终于算出了圆周率的值就在

3.1415926与3.1415927之间,准确到小数点后7位,创造了当时世界上的最高水平。直到一千多年后,德国数学家才得出相同的结果。

A: 爸爸,我们也去你的书房画个一丈的大圆吧,我们也算一算圆周率好不好?

B: 恐怕我的书房没有那么大,我们先剪一个小圆试试怎么样?

A: 好的,现在就开始吧。

生词注解 Notes

① iconic /aɪˈkɒnɪk/ *adj.* 偶像的;符号的

② verse /vɜːs/ *vt.* 使……熟练;精通

③ combination /ˌkɒmbɪˈneɪʃn/ *n.* 结合;组合

④ approximation /əˌprɒksɪˈmeɪʃn/ *n.* 近似值;粗略估算值

⑤ fraction /ˈfrækʃn/ *n.* 分数;小数

⑥ cyclotomy /saɪˈklɒtəmɪ/ *n.* 割圆术;睫状肌切开术

⑦ polygon /ˈpɒlɪɡən/ *n.* 多边形;多角形

⑧ circumference /səˈkʌmfərəns/ *n.* 圆周;圆周长

⑨ hexagon /ˈheksəɡən/ *n.* 六边形;六角形

《九章算术》

The Nine Chapters on the Art of Mathematics

导入语 Lead-in

《九章算术》是中国古典数学的重要著作之一，全面总结了战国时期、秦朝和汉朝的数学成就，共收录数学问题246个，历来被尊为"算经之首"。由西汉时期张苍和耿寿昌增补与整理，魏晋时期杰出数学家刘徽为其作注。《九章算术》以"问题集"的形式编写，把246个数学问题按照性质进行分类，包含方田、粟米、衰分、少广、商功、均输、盈不足、方程和勾股等九大类别，在几千年前就创立了中国古代数学的完整体系，是当时世界上最简练有效的应用数学，为中国乃至世界数学的发展做出了巨大贡献。

文化剪影　Cultural Outline

The Nine Chapters on the Art of Mathematics was a comprehensive work that summarized traditional mathematical thoughts from the Qin Dynasty to the Han Dynasty in China. It was a mathematical system formed by the use of arithmetic as a special method in ancient China, which constituted the rudimentary model of ancient Chinese mathematics.

《九章算术》是对我国秦代到汉代传统的数学思想进行总结的一部综合性著作,是中国古代用算筹为特殊的计算工具进而形成的一种数学体系,构成了中国古代数学文化体系的初步模型。

The Nine Chapters on the Art of Mathematics defined many important mathematical concepts, first proposed the concepts of **equations**① and fractions, discussed the problems of **simultaneous**② linear equations, fractional reduction, general division and fractional operations and put forward their solutions, laid the theoretical foundation for traditional Chinese mathematics, and completed the theoretical system of traditional Chinese mathematics.

《九章算术》界定了很多重要的数学概念,最早提出了方程和分数的概念,总结和说明了如何解答联立一次方程,系统论述了分数的约分、通分和分数运算的问题,奠定了中国传统数学的理论基础,完善了中国传统数学的理论体系。

The Nine Chapters on the Art of Mathematics embodied the practical style of algorithmic centricity in traditional Chinese mathematics, employed various of grains such as barley, millet and corn to explain the **conversion**③ of units, and the cattle, horses, and grains to **illustrate**④ the distribution problem, and reflected a strong social mathematical cultural style.

《九章算术》体现了中国古代数学以算法为中心的实用性风格,运用舂米、谷子、粟米等多种粮食来说明单位的换算,运用牛、马、粮食来说明分配问题,体现了强烈的社会性数学文化风格。

佳句点睛 Punchlines

1. *The Nine Chapters on the Art of Mathematics* has established the basic **framework**⑤ for the Chinese and Oriental mathematics.

《九章算术》确立了中国乃至东方数学的基本框架。

2. *The Nine Chapters on the Art of Mathematics*, one of the most important writings of traditional Chinese mathematics, has profoundly influenced the development of Chinese mathematics and the formation of Chinese mathematics system.

《九章算术》是中国古代数学史上的重要著作之一,深刻影响了中国数学的发展和中国数学体系的形成。

3. The status of *The Nine Chapters on the Art of Mathematics* in

the Oriental Mathematics is roughly equivalent to Euclid's *Elements of Geometry* in the Occidental mathematics. The two works are like two dazzling pearls shining in the ancient East and West.

《九章算术》在东方数学中的地位大致相当于欧几里得的《几何原本》在西方数学中的地位,这两部著作犹如两颗璀璨的明珠,在古代的东方和西方熠熠生辉。

情景对话 Situational Dialogue

A: Today, Professor Wang's analysis on *The Nine Chapters on the Art of Mathematics* gave me a new understanding of traditional Chinese mathematics.

B: What do you mean?

A: In the past, I always thought that ancient Chinese people were good at **metaphysics**① without rigorous mathematical thinking.

B: What about your ideas now?

A: Through Professor Wang's lecture, I deeply recognize that the development of mathematical thought in China is different from that in the West. *The Nine Chapters on the Art of Mathematics*, employing the examples in real life, solves the practical problems, which met the social needs in ancient China.

B: What about the contents?

A: Oh, it is divided into nine chapters, including Fangtian (rectangular fields), Sumi (millet and rice), Cuifen (proportional

distribution[7]), Shaoguang (lesser breadth), Shanggong (consultations on works), Junshu (equitable taxation), Yingbuzu (excess and **deficit**[8]), Fangcheng (**rectangular**[9] array), and Gougu (base and altitude).

B: From its contents, it is obvious that it embodies the strong social mathematical characteristics.

A: Yes. Our teacher also asked us to study it thoroughly, not only to study its content, but also its ideas and influences.

B: Come on! As a liberal arts student, I really admire the science students like you.

A: 今天，王教授对《九章算术》的解析使我对我国数学有了新的认识。

B: 此话何意？

A: 以前我一直认为我国古人擅长玄学，缺乏严谨的数学思维方式。

B: 你现在有何看法呢？

A: 通过王教授的讲解，我深刻认识到我国数学思想的发展和西方数学思想发展的体系不同，采用实际生活中的例子，注重解决实际问题，适应了古代中国的社会需求。

B:《九章算术》都有哪些内容？

A: 噢，它分为九章，包含方田、粟米、衰分、少广、商功、均输、盈不足、方程和勾股等类别。

B: 从内容显然可以看出，《九章算术》体现了强烈的社会性数学文化风格。

A: 是的。老师也要求深入研读《九章算术》,不仅研究其内容,而且研究其思想和影响。

B: 加油！作为一名文科生,真的很羡慕你们这些理科生。

生词注解　Notes

① equation /ɪˈkweɪʒn/　*n.* 方程式；等式

② simultaneous /ˌsɪmlˈteɪniəs/　*adj.* 联立的；同时发生的

③ conversion /kənˈvɜːʃn/　*n.* 转换；变换

④ illustrate /ˈɪləstreɪt/　*vt.* 阐明；图解

⑤ framework /ˈfreɪmwɜːk/　*n.* 框架；结构

⑥ metaphysics /ˌmetəˈfɪzɪks/　*adj.* 形而上学；玄学

⑦ distribution /ˌdɪstrɪˈbjuːʃn/　*n.* 分布；分配

⑧ deficit /ˈdefɪsɪt/　*n.* 赤字；不足额

⑨ rectangular / rekˈtæŋɡjələ(r)/　*adj.* 长方形的

《齐民要术》

The Important Arts for the People's Welfare

  导入语　Lead-in

人类从采集渔猎经济过渡到农业经济之后，农产品逐渐成为人类生存的主要资源。"民以食为天、国以农为本"的思想已经成为中国古人的共识。北魏时期著名农学家贾思勰编著的《齐民要术》共十卷九十二篇，位居中国古代五大农书之首，是中国目前现存最早、最完整的一部综合性农学专著，对6世纪以前中国北方地区农业生产经验进行了系统总结，重视对农业生产、科学技术和经济效益的综合性分析，强调多种经营的可行性，从而建立了比较完备的农业科学体系，所以历代统治者视其为务本劝农的治世圭臬，被誉为"中国古代农业百科全书"。

文化剪影 Cultural Outline

The Important Arts for the People's Welfare, written by Jia Sixie, an **agronomist**[①] in the Northern Wei Dynasty, covered a wide range of topics including agricultural production experiences and methods for farming, **horticulture**[②], forestry, animal husbandry, fisheries and other sideline occupations, which made the Chinese agronomy form the complete system of intensive cultivation for the first time.

北魏农学家贾思勰撰写的《齐民要术》内容丰富,涉及面广,包括农、林、牧、副、渔等方面的生产技术知识,使中国农学第一次形成了精耕细作的完整结构体系。

As a big agricultural country, China's development is inseparable from the agricultural economy. In the book *The Important Arts for the People's Welfare*, Jia Sixie stated the theory of "valuing grains that the jewelry of pearl, jade, gold and silver cannot be **edible**[③] when one is hungry and cannot be clothed when one is cold. While without the grains of millet and corn, cloth and silk in a day, one is hungry and cold. That is the reason why the wise emperors value the grains and devalue the jewelry." This theory fully embodies the idea that food is the primary **issue**[④] in the governance.

作为农业大国,中国的发展离不开农业经济。《齐民要术》所陈述的贵粟理论,"珠玉金银,饥不可食,寒不可衣;粟米布帛,一日不得而饥寒至,是故明君贵五谷而贱金玉",充分体现了"食为政首"的农为国本思想。

Through a systematic summary of the experience on farming and animal husbandry and the methods of food processing and storage, wild plant **utilization**⑤ and wasteland cultivation adopted by the working people in the middle and lower reaches of the Yellow River before the sixth century, it is clear that the book *The Important Arts for the People's Welfare* not only covers rich agronomic ideas, but also fully embodies the wisdom in livelihood and production of the Chinese ancients.

通过对6世纪以前黄河中下游地区劳动人民农牧业经验、食品加工与贮藏、野生植物利用,以及治荒方法的系统总结,《齐民要术》不仅蕴含了丰富的农学思想,而且充分体现了中国古代劳动人民的生活和生产智慧。

佳句点睛 Punchlines

1. *The Important Arts for the People's Welfare*, which covered the knowledge of agriculture, forestry, animal husbandry, fishery and other sideline occupations, was the earliest in existence and the most comprehensive **monograph**⑥ on agriculture in China.

《齐民要术》包括农、林、牧、副、渔的专业技术知识,是我国现存最早、最完整的综合性农学专著。

2. Establishing its relatively complete system of agricultural science, the book *The Important Arts for the People's Welfare* was honored as the **encyclopedia**⑦ on agriculture of ancient China.

《齐民要术》建立了相对完整的农业科学体系,被誉为"中国古代农业百科全书"。

3. *The Important Arts for the People's Welfare* fully embodied the ideas that "bread is the stall of life and agriculture is the basis of a nation".

《齐民要术》充分体现了"民以食为天、国以农为本"的思想。

 情景对话 Situational Dialogue

A: Today, we'll talk about the topic on agroecology. In modern agriculture, the application of the chemicals such as **pesticides**⑧ and the chemical fertilizers has greatly improved agricultural products. But it also caused huge damage to agroecology. Now, let's discuss the agroecological ideas contained in *The Important Arts for the People's Welfare* at first.

B: *The Important Arts for the People's Welfare* first reflects the idea that agricultural production should conform to the times. Jia Sixie believed that agricultural production should follow the seasons. As is said in the book, "In spring, the farmers should plow the land; in summer, cultivate the crops; and in autumn, harvest and store the crops". That is to say, the agricultural work on cultivation, weeding, harvesting and storage should be carried out in accordance with the seasons, so as to ensure the yield and harvest.

A: Good. Do you have any other ideas?

C: *The Important Arts for the People's Welfare* also reflects the idea that agricultural production should tailor to local conditions. Jia Sixie believed that in addition to following the times, agricultural production should accord with the geographical conditions. As is said in the book, "In agricultural production, only when farmers follow the times and the geographical conditions, can they gain more with less efforts".

A: Yes, the land conditions directly influence the growth of crops. Any more?

D: Besides following the times and geographical conditions, we should fully play the **dynamic**⑨ role of human being in agricultural production. Jia Sixie believed that the farmers played a very important role and work as a key element in agricultural production. He put forward some methods on improving agricultural cultivation, such as the soil improvement and the green manure crop rotation technique, which demonstrated his point.

A: Yes. Your analysis on the agricultural ideas reflected in *The Important Arts for the People's Welfare* is thorough. *The Important Arts for the People's Welfare* embodies the ideas on harmonious development in traditional agriculture. We should draw the wisdom from these ideas and cultivate the new concept on harmony between man and nature in modern agriculture.

A：今天我们来谈一谈农业生态这个话题。在现代农业生产中，农药、化肥等化学制品的应用大幅提升了农业产品，但农业生态也因此遭到巨大破坏。现在，我们来讨论一下《齐民要术》这部农书所包含的农业生态思想。

B：《齐民要术》首先反映了农业生产要顺应天时的农业思想。贾思勰认为，农业生产要遵循季节规律，"民春以力耕，夏以强耘，秋以收敛"，即农业耕种、除草、收割、储藏要按照季节进行，才能保证农业丰产、丰收。

A：很好。还有吗？

C：《齐民要术》还反映了因地制宜的农业生产思想。贾思勰认为，农业生产除了顺应天时之外，还要裁量地理，根据土地特点进行适合的农业生产活动，"顺天时，量地利，则用力少而成功多"。

A：是的，土地的特性直接影响农作物的生长。还有吗？

D：在顺应天时和地利的基础上，还要在农业生产中发挥人的主观能动性。贾思勰认为，人在农业生产中扮演着极其重要的角色，是一种关键要素。他提出的各种改进农业耕作的方法，如土壤的养护和绿肥轮作法，充分表明了这一点。

A：是的，同学们对《齐民要术》所反映的农业思想的分析非常透彻。纵观《齐民要术》全篇，处处蕴含着和谐共生的传统农业生态文明理念，我们应从这些理念中汲取智慧，树立人与自然和谐共生的现代农业生态文明理念。

 生词注解 Notes

① agronomist /əˈɡrɒnəmɪst/ n. 农学家

② horticulture /ˈhɔːtɪkʌltʃə(r)/ n. 园艺学；园艺

③ edible /ˈedəbl/ adj. 适宜食用的；(无毒)可以吃的

④ issue /ˈɪʃuː/ n. 重要议题；争论的问题

⑤ utilization /ˌjuːtəlaɪˈzeɪʃn/ n. 利用；应用

⑥ monograph /ˈmɒnəɡrɑːf/ n. 专著；专题文章

⑦ encyclopedia /ɪnˌsaɪkləˈpiːdiə/ n. 百科全书；(某一学科的)专科全书

⑧ pesticide /ˈpestɪsaɪd/ n. 杀虫剂；除害药物

⑨ dynamic /daɪˈnæmɪk/ adj. 动态的；发展变化的

《天工开物》

An Encyclopedia of Technology

 导入语 Lead-in

《天工开物》为明代科学家宋应星所著，共三卷十八篇，图文并茂，上卷记载五谷棉麻种植、蚕蜂饲养、纺织印染、矿产采集和采盐制糖；中卷记载砖瓦陶瓷制作、车船制造、煤炭硫磺开采以及榨油造纸方法等工艺流程；下卷记述金属矿物开采冶炼、兵器制造、颜料酒曲生产、珠玉采集加工等。《天工开物》全面总结了中国古代农业、手工业等方面卓越的科技成果，是世界上第一部关于农业和手工业生产的综合性著作，被称为"17世纪中国的工艺百科全书"。

文化剪影 Cultural Outline

An Encyclopedia of Technology was written by Song Yingxing, a scientist in the Ming Dynasty, in the form of pictures and texts, which systematically summarized the technology and achievements in agricultural and **handicraft**[①] industry in ancient China, forming a complete scientific and technological system.

明朝科学家宋应星撰写的《天工开物》,以图文并茂的形式系统总结了中国古代农业、手工业的技术和成就,构成了一个完整的科学技术体系。

An Encyclopedia of Technology, employing a large number of concepts and quantitative description of technical data, objectively described the advanced scientific and technological achievements of agriculture and handcraft industry in ancient China, reflecting the advanced research methods. Meanwhile, it **emphasized**[②] some scientific spirits such as harmony between human and nature, **coordination**[③] between human power and natural forces. It held the point that human beings should exploit the natural resources by technology to meet the needs of their material life and spiritual life, and that human being should **exert**[④] their initiative in their production work.

《天工开物》引用大量的理论概念和对技术数据的定量描述,客观生动地记述了中国古代农业和手工业的先进科技成果,体现了先进的研究方法。同时,《天工开物》强调了人与自然相协调、人力与自然力相配合的科学精神,通过技术从自然资源中开发物产,以满足人的物质生活和精神生活的需要,从而使人在自然界面前发挥主观能动性。

In the order of "valuing the grains and devaluing the jewelry", *An Encyclopedia of Technology* detailed the aspects in food, clothing, housing and transportation. It not only reflected the situation of productive force as **capitalism**⑥ **sprouted**⑥ in the late Ming Dynasty, but also **facilitated**⑦ the progress of science and technology in the world.

《天工开物》按照"重五谷而贱金玉"的顺序,对衣食住行进行了详细记载和描绘,不仅反映了明朝末年中国出现的资本主义萌芽的生产力状况,而且推动了世界科技的进步。

佳句点睛 Punchlines

1. *An Encyclopedia of Technology* was the first comprehensive work on agriculture and handicraft industry in the world.

《天工开物》是世界上第一部关于农业和手工业生产的综合性著作。

2. *An Encyclopedia of Technology* emphasized the scientific spirit of harmony between human and nature as well as coordination between human power and natural forces.

科技成就

《天工开物》强调了人与自然相协调、人力与自然力相配合的科学精神。

3. *An Encyclopedia of Technology* was honored as "The Chinese technology encyclopedia in the 17th century".

《天工开物》被誉为"17世纪中国的工艺百科全书"。

 情景对话 **Situational Dialogue**

A: I want to write my thesis on *An Encyclopedia of Technology*. Can you give me some tips?

B: Of course. First, you should share your ideas with me.

A: As a landmark masterpiece in the history of science and technology in China, *An Encyclopedia of Technology* has some **insights**① for modern ecological civilization.

B: You have a good idea. But you should fully realize the difficulties you will encounter on this topic.

A: I know. As a work of the Ming Dynasty, *An Encyclopedia of Technology* was written in classical Chinese. To understand this book requires knowledge of ancient Chinese, as well as scientific and technical knowledge, including physics, chemistry, agronomy, and so on. Don't worry. Although I'm majoring in engineering, I was quite good at Chinese in high school, especially in classical Chinese.

B: Well, that's a relief. Citing a number of theoretical concepts and quantitative descriptions of technical data, *An Encyclopedia of Technology* objectively and vividly described the advanced scientific and technological achievements of ancient Chinese agriculture and handicraft industry. You should consult a lot of literature and professionals if necessary, and carefully sorted out the **ecological**[①] philosophy, ecological balance view and ecological economy view.

A: OK, with your encouragement and support, I have much more confidence in my thesis writing.

B: Come on! Please contact me when you need my help.

A: 我想写一篇关于《天工开物》的论文。你能给我一些建议吗？

B: 当然可以。首先，你可以分享一下你的想法。

A: 作为我国古代科技史上里程碑式的巨作，《天工开物》反映的生态思想对现代生态文明建设具有一定的借鉴意义。

B: 你的想法很好，但你要充分认识到写这个选题的论文将会遇到的困难。

A: 我知道。《天工开物》是明代的著作，看懂这本书首先需要古文知识，同时还要具备科技知识，包括物理、化学、农学知识等。放心，虽然我学的是工科，但我高中时语文学得相当好，特别喜欢文言文。

B: 这我就放心了。《天工开物》引用了大量的理论概念和对技术数据的定量描述，客观生动地记述了中国古代农业和手工业的先进科技成果，你要多方查阅文献，必要时可以请教专业人士，认真梳理其所体现的生态哲学观、生态平衡观和生态经济观等科技成就思想。

A：好的，有你的鼓励和支持，我对论文写作就更有信心了。

B：加油！如需帮忙可随时联系我。

生词注解 Notes

① handicraft /ˈhændɪkrɑːft/ *n.* 手工艺；手工艺品

② emphasize /ˈemfəsaɪz/ *vt.* 强调；重视

③ coordination /kəʊˌɔːdɪˈneɪʃn/ *n.* 协调；协作

④ exert /ɪɡˈzɜːt/ *vt.* 运用；行使

⑤ capitalism /ˈkæpɪtəlɪzəm/ *n.* 资本主义

⑥ sprout /spraʊt/ *vt.* 萌芽；发芽

⑦ facilitate /fəˈsɪlɪteɪt/ *vt.* 促进；促使

⑧ insight /ˈɪnsaɪt/ *n.* 领悟；洞察力

⑨ ecological /ˌiːkəˈlɒdʒɪkl/ *adj.* 生态的；生态学的

赵州桥

Zhaozhou Bridge

导入语 Lead-in

赵州桥原名安济桥,坐落在河北省石家庄市赵县的洨河上,由隋朝著名匠师李春设计和制造,是当今世界上建造最早、保存最完善的古代敞肩石拱桥。赵州桥的设计合乎科学原理,造型优美、线条流畅、石雕丰富,在建筑技术、结构设计和装饰艺术上都达到了中国古代桥梁建筑的高峰。1961年被国务院列为第一批全国重点文物保护单位。1991年,赵州桥被美国土木工程师学会选定为世界第12处"国际土木工程历史古迹",和英国伦敦的塔桥、法国巴黎的埃菲尔铁塔等建筑一样,成为世界历史上最辉煌的土木工程典范。

科技成就

文化剪影　Cultural Outline

Zhaozhou Bridge, built in the Sui Dynasty, is 64.4 meters long, 37.02 meters wide and 7.32 meters high. It is the oldest and best preserved single-**span**① open-shouldered stone arch bridge in the world.

赵州桥建于隋朝,桥长64.4米,跨径37.02米,桥高7.32米,是当今世界上建造最早、保存最完善的单孔敞肩式石拱桥。

The span of the single-hole arch increases the flood discharge **capacity**② and the safety of Zhaozhou Bridge, which is the pioneering undertaking in the history of bridge construction in China. The **arc**③ of its big single-hole stone arch is less than that of the semicircle. And there are four smaller arches on the left and right exterior curve of the main arch, which gives the bridge a light and elegant **profile**④ with a beautiful curved line of the big arch complemented by four smaller symmetrical arches.

赵州桥单孔长跨的形式增加了石桥的泄洪能力和安全性,是中国桥梁史上空前的创举。其单孔大石拱是小于半圆的一段弧形,在大石拱的双肩上各有两个圆弧形的小石拱,四个小拱呈现均衡对称设计,大拱与小拱构成了一幅完整的图画,桥身显得更加娟秀轻盈。

The abundant stone carvings, such as the **coiling**⑤ dragons, **squatting**⑥ lions, sucking beasts, flowers and bamboos, make Zhaozhou Bridge a fine work of art, not only highlighting the **exquisite**⑦ skills of the carving art in ancient China, but also reflecting the profound traditional culture in China.

盘龙、蹲狮、吸水兽、花饰、竹节等丰富的石雕使赵州桥成为一件精美的艺术品,不仅彰显了中国古代精湛的雕刻技艺,而且体现了中国博大的传统文化。

佳句点睛 Punchlines

1. Zhaozhou Bridge, spanning the river is like a crescent moon rising above the clouds or a long rainbow rising after the rain.

横跨水面的赵州桥,犹如初出云层的一轮新月,又像是雨后初霁的一道长虹。

2. Zhaozhou Bridge has reached the peak of the bridge construction in ancient China in terms of **architectural**⑧ technology, structural design and decorative art.

赵州桥在建筑技术、结构设计和装饰艺术方面都达到了中国古代桥梁建筑的高峰。

3. The carved designs on the jade railing are exquisite, reflecting the wisdom and superb skills of the craftsmen in ancient China.

玉石栏杆上雕刻的图案精美细致,栩栩如生,体现了中国古代工匠的聪明才智和高超技艺。

情景对话 **Situational Dialogue**

A: Shall we take pictures here?

B: Of course. The scenery here is so **gorgeous**①. We use Zhaozhou Bridge as the background.

A: Are you OK?

B: OK. Let's take a walk on the bridge.

A: Here is the instruction of Zhaozhou Bridge. It has a history of 1,400 years. And it is the oldest open-spandrel stone arch bridge best preserved in China and in the world.

B: Yes. It was built by Li Chun, a famous craftsman in the Sui Dynasty.

A: Wow! Please look at these dragon patterns on the railings. Some are intertwined with their forepaws touching each other, some playing a ball. All the dragons seem to be swimming as if they were alive.

B: Besides the coiling dragons, there're still some other beautiful stone carvings. Please look at these flowers and bamboos. These exquisite carvings make Zhaozhou Bridge a fine work of art.

A: Really great! Let's look at the structure of the bridge from the bank.

B: Okay.

A: 我们就在这里拍照好吗?

B: 当然可以,这里风景很好,我们可以以赵州桥为背景。

A: 好了吗?

B: 好了。我们去桥上走走吧。

A: 这里有赵州桥的介绍,赵州桥距今已有一千四百多年的历史了,是当今中国乃至世界上最古老、保存最为完善的敞肩圆弧石拱桥。

B: 是的,赵州桥是隋朝著名工匠李春修建的。

A: 哇,你看这栏板上的各种盘龙图案,有两条相互缠绕的龙,前爪相互抵着,各自回首遥望,还有双龙戏珠,所有的龙似乎都在游动,栩栩如生。

B: 是的,除了盘龙,还有其他一些漂亮雕饰。请看这些花饰和竹节。这些独具匠心的雕刻使赵州桥成为了一件精美艺术品。

A: 真了不起。我们去河岸上看看桥的结构吧。

B: 好的。

生词注解 Notes

① span /spæn/ n.(桥或拱的)墩距;跨度

② capacity /kəˈpæsətɪ/ n. 能力;地位

③ arc /ɑːk/ n. 弧形物;弧度

④ profile /ˈprəʊfaɪl/ n. 侧面轮廓;形象

⑤ coil /kɔɪl/ vt. 缠绕;盘绕

⑥ squat /skwɒt/　*v.* 蹲坐；蹲

⑦ exquisite /ɪkˈskwɪzɪt/　*adj.* 精美的；精致的

⑧ architectural /ˌɑːkɪˈtektʃərəl/　*adj.* 建筑学的；建筑上的

⑨ gorgeous /ˈɡɔːdʒəs/　*adj.* 极好的；灿烂的

都江堰

Dujiangyan

导入语 Lead-in

都江堰是四川成都平原上一个集防洪、灌溉、航运于一体的综合水利工程,是迄今为止世界上年代最久、唯一留存、以无坝引水为特征、科学完整的庞大水利工程体系,被誉为"世界水利文化的鼻祖"。在秦朝、三国时期和唐朝,根据所处地理位置和筑堤方法,都江堰分别被称为"湔堋""都安堰"和"楗尾堰",直到宋朝才首次提及"都江堰"并用其来准确概括整个水利系统工程,一直沿用至今。都江堰水利工程的建设,使成都平原成为"天府之国",变害为利,人、地、水三者达到了高度和谐统一,开创了中国水利史上的新纪元。2000年被联合国教科文组织列入《世界文化遗产名录》,2020年入选"巴蜀文化旅游走廊新地标"。

文化剪影 Cultural Outline

Dujiangyan **Irrigation**[①] System was built by Li Bing, prefect of Shu for the Qin State, and his son, who learned from the previous experience of flood control and made full use of the **naturalization**[②] of the mountain and the river, and spent eight years in completing the project with the local people. It has effectively controlled the floods in the Minjiang River, making the Chengdu Plains a land of **abundance**[③].

秦国蜀郡太守李冰及其儿子吸取前人的治水经验，充分利用山势和水体的自然造化，率领当地人民历经八年努力，建成了都江堰水利工程，该工程有效治理了岷江水患，使成都平原成为水旱从人、沃野千里的天府之国。

Dujiangyan is a colossal, comprehensive and scientific **hydraulic**[④] project. Its three main constructions as the Yuzui Diversion Dike, Feishayan Spillway and Baopingkou Intake work interdependently, solving the problems of water flow diversion, sand discharge and water inflow control to ensure against the flood and **draught**[⑤] caused by the Minjiang River, whose bed is apparently higher than the land of the Chengdu Plains. It is praised as the "**unparalleled**[⑥] irrigation method in the world".

都江堰是一个科学完整的庞大水利工程体系，其鱼嘴分水堤、飞沙堰泄洪道和宝瓶口引水口等主体工程相互依赖、功能互补，形成布局合理的系统工程,科学解决了江水自动分流、自动排沙、控制进水流量等问题,消除了岷江在成都平原上悬江的水患,变水害为水利,被誉为"世界上无与伦比的灌溉方法"。

The construction of Dujiangyan Irrigation System embodied the Chinese ancients' industry, courage and wisdom, from the choice of water diversion **outlet**① location of the Minjiang River to the practice of burning the stones of the mountain to divert the river and to the method of piping up bamboo cages with pebbles to **regulate**② the water volume.

从岷江分水口地理位置的选择,到以火烧石凿山引水,再到竹笼装卵石堆筑以调节水量的办法,都江堰水利工程的修建无不体现了中国古代劳动人民的勤劳、勇敢和智慧。

佳句点睛 Punchlines

1. Dujiangyan is a comprehensive water conservancy project for flood control, irrigation and shipping on the Chengdu Plains of Sichuan.
都江堰是四川成都平原上一个集防洪、灌溉、航运于一体的综合水利工程。

2. Dujiangyan is characterized by diversion without any dam, which is the only great water conservancy project in the world so far.
都江堰以无坝引水为特征,是全世界迄今为止仅存的一项宏大的水利工程。

3. The construction of Dujiangyan Irrigation System embodied the Chinese ancients' industry, bravery and wisdom.

都江堰水利工程的修建体现了中国古代劳动人民的勤劳、勇敢和智慧。

情景对话 Situational Dialogue

A: I heard you visited Dujiangyan in the summer vacation?

B: Yes, it is awesome. The scenery is not only beautiful, but the cultural and historical significance behind it is even more remarkable.

A: Really? Could you tell me something about it?

B: Of course. Dujiangyan, first built in the last years of King Zhao of Qin, is a colossal, comprehensive and scientific hydraulic project. Such a great hydraulic project was unique in the world over two thousand years ago. Even nowadays, it is still one of the great projects in water conservancy engineering in the world.

A: Wow, amazing. Who built it?

B: It was built by Li Bing, prefect of Shu for the Qin State, and his son, who learned from the previous experience of flood control and made full use of the naturalization of the mountain and the river, and spent eight years in completing the project with the local people.

A: Too great. It's still working after more than two thousand years.

B: I agree with you. This water conservancy project benefits the local people in the past, at present and even in the future. It is a model of regional water conversancy projects.

A: Thank you. I've learned a lot about Dujiangyan. I can't wait to see it.

B: Alright, my pleasure.

A: 听说你暑假游览了都江堰?

B: 是的,那地方太棒了,不仅风景秀美,其背后的文化历史意义更是了不起。

A: 真的吗? 你能给我讲讲吗?

B: 当然可以,都江堰始建于秦昭王末年,是一个科学、完整的庞大水利工程体系。两千多年前的都江堰可以取得这样的成就,这在世界上都是绝无仅有的,都江堰至今仍是世界上伟大的水利工程之一。

A: 哇,太了不起了。这是谁修建的?

B: 秦国蜀郡太守李冰及其儿子吸取前人的治水经验,充分利用山势和水体的自然造化,率领当地人民历经八年的努力修建而成。

A: 太伟大了。两千多年之后都江堰还可以使用。

B: 是的,李冰修建的都江堰是造福古今、惠泽未来的水利工程,是区域水利网络化的典范。

A: 谢谢你。我了解了很多关于都江堰的知识。我都迫不及待想去看一看了。

B: 好的,不客气。

生词注解 Notes

① irrigation /ˌɪrɪˈɡeɪʃn/ n. 灌溉；灌水

② naturalization /ˌnætʃrəlaɪˈzeɪʃn/ n. 自然化；归化

③ abundance /əˈbʌndəns/ n. 充裕；丰富

④ hydraulic /haɪˈdrɒlɪk/ adj. 水力的；水力学的

⑤ draught /drɑːft/ n. 气流；汇票

⑥ unparalleled /ʌnˈpærəleld/ adj. 无与伦比的；空前的

⑦ outlet /ˈaʊtlet/ n. 出水口；排气口

⑧ regulate /ˈreɡjuleɪt/ vt. 调节；控制

古代科技成就 第一部分

《墨经》

The Mohist Canon

导入语 Lead-in

《墨子》是战国时期墨家著作的总集，由墨翟及其弟子所著。《墨经》是《墨子》的重要部分，约完成于前388年。《墨经》也称《墨辩》，包括《经上》《经下》《经上说》《经下说》《大取》和《小取》六篇，主要讨论认识论、逻辑和自然科学方面的问题。《墨经》还论述了力学、光学、几何学、工程技术、现代物理学和数学等基本要素，并阐述了影、小孔成像、平面镜、凹面镜、凸面镜成像的原理，以及焦距和物体成像的关系，这些比古希腊欧几里得的光学记载早了一百多年。力学方面的论说也是古代力学的代表作，对力的定义、杠杆、滑轮、轮轴、斜面及物体沉浮、平衡和重心都有论述。

文化剪影 Cultural Outline

The Mohist Canon was an ancient Chinese work from the Warring States Period that **expounded**① the philosophy of Mohism. *The Mohist Canon* was a pioneer in the study of Chinese logic as well as in that of metaphysics, mechanics, optics and geometry.

《墨经》是一部中国古代的著作,写于战国时期,阐述了墨家哲学。《墨经》是中国逻辑学研究的先驱,也是形而上学、力学、光学和几何学方面研究的先驱。

The Mohist Canon, also known as "*Mobian*", mainly discussed the issues of epistemology, logic and natural science.

《墨经》,亦称《墨辩》,主要讨论认识论、逻辑和自然科学方面的问题。

The Mohist Canon was one of the important documents in the history of Chinese **ethics**②, philosophy of language, logic and science, and an important record of the scientific achievements of the pre-Qin Era.

《墨经》是中国伦理学、语言哲学、逻辑学和科学史上重要的著作之一,是对先秦时代科学成就的重要记录。

佳句点睛 Punchlines

1. Mo Di was a handicraftsman who made machinery and was proficient in carpentry and enthusiastic in the research of natural sciences.

墨翟是一位制造机械的手工业者,他精通木工,同时热心研究自然科学。

2. *The Mohist Canon* recorded the discoveries of diverse fields such as geometry, mechanics, **optics**③ and economics.

《墨经》记载了几何学、力学、光学和经济学等不同领域的发现。

3. *The Mohist Canon* pointed out that shadows are caused by objects obstructing the movement of light. When the object that obstructs the light moves, the shadow appears to be moving on the surface. In fact, this is a physical process where the original shadow disappears and new shadows continue to form.

《墨经》指出,影子是物体阻碍光线的行进而造成的。当阻碍光线的物体移动时,表面看来影子也在移动,实际上这是原影不断消失、新影不断形成的物理过程。

情景对话 Situational Dialogue

A: Lucy, do you know why was China's quantum communication

satellite nicknamed as "Mozi"?

B: Yes, it's **in homage to**① the philosopher and his writings on optics.

A: Great! Mozi and his students made the world's first experiment of **inverted**⑤ images of small holes, pointed out the nature of light along a straight line, and summarized similar theories more than two thousand years before Newton.

B: Did *The Mohist Canon* record all his achievements?

A: Yes, and it also provided the definitions of circumference, diameter, radius and volume.

B: It is said that Mozi was a carpenter, extremely skilled in creating devices.

A: Yes. Though he did not hold a high official position, Mozi was sought out by various rulers as an expert on fortification.

B: In this way, he was similar to the other knights-errant of the period.

A: Furthermore, he emphasized universal love. He travelled from one crisis zone to another throughout the **ravaged**⑥ land of the Warring States Period, trying to **dissuade**⑦ rulers from their plans of conquest.

B: He was best known for his idea of universal love.

A: 露西，你知道为什么中国的量子通信卫星叫"墨子号"吗？
B: 知道，是为了向这位哲学家及其有关光学的著作致敬。
A: 太棒了！墨子和他的学生做了世界上第一个关于小孔成像

的实验,指出了光沿直线的性质,比牛顿得出这个结论早了两千多年。

B:《墨经》中记录了他的这些成就吗?

A:是的,书中还给周长、直径、半径和体积作出了定义。

B:据说墨子是名木匠,非常擅长制作器具。

A:是的。虽然墨子没有很高的官职,但许多统治者将他视作防御工事的专家,四处寻访他。

B:这么看来,他和当时那些游侠骑士相似。

A:而且,墨子强调兼爱。他在危机地带间游走,穿越战国时期满目疮痍的大地,试图劝阻统治者放弃征战的计划。

B:墨子最出名的就是他的兼爱说。

生词注解 Notes

① expound /ɪkˈspaʊnd/ vt. 详述;阐述

② ethic /ˈeθɪk/ n. 伦理;道德规范

③ optics /ˈɒptɪks/ n. 光学

④ in homage to 向……致敬

⑤ invert /ɪnˈvɜːt/ vt. 使……颠倒;使……反转

⑥ ravage /ˈrævɪdʒ/ vt. 毁坏;掠夺

⑦ dissuade /dɪˈsweɪd/ vt. 劝阻;劝止

《营造法式》

The Rules of Architecture

导入语 Lead-in

《营造法式》是中国第一本详细论述建筑工程做法的官方著作,目的是为了控制官式建筑的用料与用工,以节省政府的财政开支。全书现存34卷,357篇,由宋代著名建筑学家李诫在两浙工匠喻皓《木经》的基础上编修宋代建筑的做法、用工、图样等资料,规定了各种建筑施工设计、用料、结构、比例等方面的标准,于1103年刊行全国。《营造法式》不仅体例较好,便于灵活应用,而且内容丰富,阐述精确,是中国古代优秀的建筑著作之一,也是了解宋代建筑的一把金钥匙。它具有极高的科学价值,在中国古代建筑史上也发挥了承前启后的作用,对后世建筑技术的进一步发展产生了极为深远的影响。

文化剪影　Cultural Outline

The Rules of Architecture was the oldest **extant**① Chinese technical manual on buildings. It was compiled by Li Jie, director of palace buildings of the Northern Song, and published in 1103.

《营造法式》是现存最古老的中国建筑技术手册，由北宋时期的宫殿建筑主管李诫编撰，并于1103年出版。

The Rules of Architecture **specified**② the units of measurement, the design standards and the construction principles, as well as the building elements **illustrated**③ in drawings.

《营造法式》规定了计量单位、设计标准和施工原则，以及附图所示的建筑元素。

The Rules of Architecture provided the **authoritative**④ guidance to leaders of public works projects, such as rigorous construction procedures and effective **budget**⑤ management.

《营造法式》为公共建设项目的负责人提供了权威的指导，比如严谨的建筑程序和有效的预算管理。

佳句点睛　Punchlines

1. The Rules of Architecture included a large number of illustra-

tions showcasing the specific details of building structures and separate parts.

《营造法式》中有大量插图，展示了建筑结构和单个部件的具体细节。

2. The complex system of structural bracketing is one of the distinctive **traits**⁶ of traditional Chinese architecture.

复杂的结构体系是中国传统建筑的显著特征之一。

3. Materials and labor are largely standardized, so it is easy to calculate how many materials and labor are needed to build a certain type of building.

材料和劳动力在很大程度上是标准化的，因此很容易计算出建造某种类型的建筑需要多少材料和劳动力。

 情景对话 **Situational Dialogue**

A: I think that architecture itself proclaims or symbolizes wisdom and virtue.

B: In fact, it was not until the Northern Song Dynasty that authoritative works on architectural standards and methods appeared.

A: Do you mean *The Rules of Architecture*?

B: Yes, it aimed to reduce waste in materials and expenses and prevent corruption in the construction process of local administrations.

A: It is said that it was the earliest and the most comprehensive Chinese **treatise**⑦ on architectural technology.

B: Yes, it **stipulated**⑧ standard amounts of materials, labor, and working time needed in executing the given construction tasks.

A: Wow, it was a remarkable project.

B: What's more, it provides hand-drawn illustrations of all the practices and standards including carving, tile decorations and color-painting **motifs**⑨.

A: Not only was it included in the imperial compilations, but private scholars and book collectors also treasured and copied it by hand generation after generation.

B: Such a complete record of building technologies and methods with rich illustrations is **unprecedented**⑩ in the human history.

A: 我觉得建筑本身就象征着智慧和美德。

B: 事实上,直到北宋时期才出现了关于建筑标准和方法的权威著作。

A: 你是说《营造法式》吧?

B: 是的,它旨在减少材料和费用的浪费,防止地方行政部门在建设过程中出现贪污现象。

A: 据说它是中国最早、最全面的建筑技术专著。

B: 是啊,它规定了执行给定的施工任务所需材料的数量标准、劳动力和工作时长。

A: 哇,真是个了不起的工程。

科技成就

B: 另外，它还提供了所有做法和标准的手绘插图，包括雕刻、瓷砖装饰和彩绘图案。

A: 它不仅被收录在皇家的汇编中，而且受到私人学者和藏书家珍视，并通过手抄本代代传承。

B: 如此完整并带有丰富插图的建筑技术和方法的记录，在人类历史上是前所未有的。

生词注解　Notes

① extant /ekˈstænt/　*adj.* 现存的；显著的

② specify /ˈspesɪfaɪ/　*vt.* 指定；详细说明

③ illustrate /ˈɪləstreɪt/　*vt.* 阐明；图解

④ authoritative /ɔːˈθɒrətətɪv/　*adj.* 权威的；(关于某一学科)权威性的

⑤ budget /ˈbʌdʒɪt/　*n.* 预算；预算费

⑥ trait /treɪt/　*n.* 特性；品质

⑦ treatise /ˈtriːtɪs/　*n.* 论述；专著

⑧ stipulate /ˈstɪpjuleɪt/　*vt.* 规定；保证

⑨ motif /məʊˈtiːf/　*n.* 图形；主旨

⑩ unprecedented /ʌnˈpresɪdentɪd/　*adj.* 空前的；史无前例的

《考工记》

The Book of Diverse Crafts

导入语 Lead-in

《考工记》是春秋战国时期记述官营手工业各工种规范和制造工艺的文献,被收录至《周礼》"冬官篇"并流传至今。全书仅7,100余字,但科技信息含量极其丰富,涉及先秦时期的制车、兵器、礼器、钟磬、练染、建筑、水利等30个工种的技术规则,记述了一系列生产管理和营建制度,反映了当时中国的科技及工艺水平。《考工记》是中国目前所见年代最早的手工业技术文献,在中国科技史、工艺美术史和文化史上都占有重要地位。此外,《考工记》还有数学、地理学、力学、声学、建筑学等多方面的知识和经验总结。《考工记》中亦记载了中国古代创制的六种铜锡比例不同的合金成分配比,是中国也是世界上最早的合金配制记载。

文化剪影　Cultural Outline

The Book of Diverse Crafts was the oldest historic document specifying China's artifacts and techniques in the Pre-Qin Period, in which various manufacturing methods, **configurations**[①] and specifications were detailed.

《考工记》是中国记载先秦时期器物和技术的最古老文献,详细描述了各种器物的制造方法、构造和规格。

The Book of Diverse Crafts described thirty techniques used at the time. The most **prominent**[②] are bronze casting, the manufacture of carriages and weapons, a metrological standard, the making of musical instruments, and the planning of cities.

《考工记》描述了当时使用的30种技术。其中最突出的是青铜铸造、马车和武器的制造、计量标准、乐器制造和城市规划。

The Book of Diverse Crafts, an official document of the Qi State during the Spring and Autumn Period, was used to guide the handicraft industries, including the examination and **assessment**[③] of craftsmen.

《考工记》是春秋时期齐国的官书,被用来指导手工业,包括对工匠的考核和评估。

佳句点睛 Punchlines

1. The description of weapons in *The Book of Diverse Crafts* was particularly comprehensive and concrete, showing that the society attached great importance to weapons at that time.

《考工记》中对武器的描写尤为全面和具体,可见当时社会对武器的重视。

2. At present, the study on historical **artifacts**④ is still focused on *The Book of Diverse Crafts*.

目前对历史器物的研究仍以《考工记》为重点。

3. *The Book of Diverse Crafts* covered carpentry, metal, leather, **porcelain**⑤ and many others, recording the manufacturing of vehicles, weapons and musical instruments, and the building of houses.

《考工记》涉及木工、金属、皮革、瓷器和其他许多行业,还记录了车辆、武器和乐器的制造,以及房屋的建造。

情景对话 Situational Dialogue

A: After listening to the lecture just now, I think *The Book of Diverse Crafts* was indeed a classical works in the history of ancient China's science and technology.

B: Yes, it detailed information on special skills such as silk spinning, dyeing, **tanning**[6], the construction of city walls, and the casting of bronze tools.

A: It also gave insight into the arts of woodcarving, the processing of metal and animal skins, and the production of **ceramics**[7].

B: I can see you're really into it.

A: Yes, I'm interested in the part of vehicles. According to the special usages, there were the war chariots, the hunting carriages and the travel carriages. What about you?

B: Of course, the weaponry. The descriptions on the manufacturing of bows and arrows were very detailed in *The Book of Diverse Crafts*.

A: And the chapter on music was quite attractive, illustrating the different types of bells and drums.

B: Yes, the determination of correct notes was related to arithmetic, showcasing the achievements of Chinese mathematicians during that age.

A: No doubt, *The Book of Diverse Crafts* had a **profound**[8] impact on China's science and technology with more than two thousand years.

B: Yes. Although many tools have fallen out of fashion, the traces they leave behind still reveal their glorious past.

A: 听了刚才的讲座，我觉得《考工记》的确是中国古代科技史上的经典著作。

B：是的，《考工记》提供了许多技艺的详细信息，如纺丝、染色、鞣制、城墙的建造以及青铜工具的铸造。

A：它还让人们深入了解了木雕艺术、金属和皮革加工，以及陶瓷的生产等。

B：我看你听得非常投入。

A：是的，我对车辆的部分很感兴趣。车辆按其特殊用途可分为战车、猎车和旅车。你对什么比较感兴趣呢？

B：当然是武器啦。《考工记》中对弓箭的制作有很详细的描述。

A：关于音乐的章节也很吸引人，它介绍了不同类型的铃、鼓和声石。

B：是的，音符的确定还与算术有关，这些都展示了那个时代中国数学家的成就。

A：毫无疑问，《考工记》对中国两千多年的科学技术产生了深远的影响。

B：是的。虽然很多工具已经退出了历史舞台，但留下的痕迹仍然显示了它们辉煌的过去。

生词注解 Notes

① configuration /kənˌfɪɡəˈreɪʃn/　*n.* 配置；结构

② prominent /ˈprɒmɪnənt/　*adj.* 显著的；卓越的

③ assessment /əˈsesmənt/　*n.* 评估；评定

④ artifact /ˈɑːtɪfækt/　*n.* 人工制品；手工艺品

⑤ porcelain /ˈpɔːsəlɪn/ n. 瓷；瓷器

⑥ tan /tæn/ v. 鞣(革)；晒成褐色

⑦ ceramic /səˈræmɪk/ n. 陶瓷；陶瓷制品

⑧ profound /prəˈfaʊnd/ adj. 深厚的；意义深远的

《梦溪笔谈》

Brush Talks from Dream Brook

导入语 Lead-in

《梦溪笔谈》是一部涉及古代中国自然科学、工艺技术和社会历史现象的综合性笔记体著作,由北宋科学家、政治家沈括所著,收录了他一生的所见所闻和个人见解,包括《笔谈》《补笔谈》《续笔谈》三部分,内容涉及天文、历法、气象、地质、地理、物理、化学、生物、农业、水利、建筑、医药等学科,详细记载了劳动人民在科学技术方面的卓越贡献和丰硕成果,反映了中国古代尤其是北宋时期自然科学取得的辉煌成就。《梦溪笔谈》中约三分之一的篇幅记述了自然科学知识,详细记载了"布衣毕昇"发明的泥活字印刷术,这是世界上最早的关于活字印刷

的可靠史料,是中国科学史上百科全书式的重要文献。

文化剪影 Cultural Outline

Brush Talks from Dream Brook was a profound book written by Shen Kuo, a famous writer and statesman in the Song Dynasty, and is recognized as a milestone in the history of Chinese science.

《梦溪笔谈》是宋代文学家、政治家沈括创作的一部博大精深的书籍,被公认为中国科学史上的一座里程碑。

Brush Talks from Dream Brook was divided into three parts, **comprising**① 507 essays, which explored a wide range of scientific topics and explains many of Shen's discoveries and theories, as well as natural and technological phenomena.

《梦溪笔谈》分为3部分,共507篇。内容涵盖了广泛的科学主题,并阐释了沈括的许多发现和理论,以及自然和技术现象。

Brush Talks from Dream Brook covered a great variety of subjects, such as **astronomy**②, physics, medicine, laws, music, mathematics, engineering, military affairs, and so on.

《梦溪笔谈》涉及的学科非常广泛,包括天文学、物理学、医学、法学、音乐、数学、工程学、军事学等。

佳句点睛 Punchlines

1. It was the first work in the world to refer to the **magnetic**③ compass, described movable type printing and explained the origin of **fossils**④.

这是世界上第一本提到磁罗盘、描述活字印刷并且解释化石起源的著作。

2. Shen Kuo, who was **versatile**⑤, detailed the movable type printing technology invented by his **contemporary**⑥ Bi Sheng.

沈括多才多艺,详细描述了与他同时代的毕昇发明的活字印刷技术。

3. He explained the rainbow as atmospheric **refraction**⑦ and was the first person to record the phenomenon of unidentified flying objects.

他将彩虹解释为大气折射,且是最早记录不明飞行物现象的人。

情景对话 Situational Dialogue

A: Do you know the author of *Brush Talks from Dream Brook*?

B: Yes, I do. It's Shen Kuo, who was a great scientist and statesman in the early Northern Song Dynasty.

A: And he was a boy of **insatiable**⑧ curiosity who took an interest in everything around him.

B: Yes, he traveled to many parts of the country and studied the plants, fossils, rivers, mountains, weather, local engineering projects and astronomical phenomena.

A: No wonder the British historian Joseph Needham praised him as one of the great scientific minds in the Chinese history.

B: In addition, he spent more than three years drawing more than two hundred sketches before he finally came up with the concept of true north. He noted that compasses pointed to the magnetic north pole, but not to true north. The rest of the world only began to know of this more than four hundred years later.

A: Why was it entitled *Brush Talks from Dream Brook*?

B: After retirement, Shen Kuo settled down in a garden estate near modern-day Zhenjiang of Jiangsu Province and named the estate "Dream Brook", where in almost total **isolation**①, he compiled this magnificent masterpiece shining through ages.

A: 你了解《梦溪笔谈》的作者吗？

B: 了解，就是沈括啊。他是北宋初期一位伟大的科学家、政治家。

A: 他从小就充满好奇心，对周围的一切都很感兴趣。

B: 是啊，他游历了全国各地，研究植物、化石、河流、山脉、天气、当地的工程项目和天文现象。

A: 难怪英国历史学家李约瑟称赞他是中国历史上伟大的科学

头脑之一。

B: 此外，他花了三年多时间画了两百多张草图，最后提出了"真北"的概念。他指出罗盘指向的是磁北极，而不是真正的北极。世界上的其他国家直到四百多年后才知道这一点。

A:《梦溪笔谈》因何得名呢？

B: 退休后，沈括在江苏镇江附近的一座花园中定居下来，并将其命名为"梦溪"。在那里，他在几乎与世隔绝的情况下编纂了这部彪炳千秋的皇皇巨著。

生词注解 Notes

① comprise /kəmˈpraɪz/　vt. 包含；由……组成

② astronomy /əˈstrɒnəmɪ/　n. 天文学

③ magnetic /mægˈnetɪk/　adj. 地磁的；有磁性的

④ fossil /ˈfɒsl/　n. 化石；僵化的事物

⑤ versatile /ˈvɜːsətaɪl/　adj. 多才多艺的；多面手的

⑥ contemporary /kənˈtemprərɪ/　n. 同时代的人

⑦ refraction /rɪˈfrækʃn/　n. 折射；折光

⑧ insatiable /ɪnˈseɪʃəbl/　adj. 贪得无厌的；不知足的

⑨ isolation /ˌaɪsəˈleɪʃn/　n. 隔离；孤立

科技成就

《徐霞客游记》

Xu Xiake's Travels

 导入语　Lead-in

《徐霞客游记》是明代地理学家徐霞客的旅游日记，记载了他三十多年游历中所经历和观察到的各种地理现象、自然规律、气候状况、动植物情况、地下热水、大小河流、地表地下岩溶、人文地理以及少数民族地区的风俗习惯、经济状况等，是地理学家和考古学家不可多得的研究材料。《徐霞客游记》内容丰富，既是一部杰出的文学游记，又是一部极富学术价值的地理名著，同时还是世界上关于喀斯特地貌学和洞穴学最早的重要科学文献，代表了当时最先进的水平。

文化剪影　Cultural Outline

Xu Xiake's Travels, written by the geographer and travel writer Xu Xiake in the Ming Dynasty, includes his travel logs and records during his thirty-four-year journey across China. He had made detailed records of geography, **hydrology**[①], geology, plants and other phenomena, and outstanding achievements in geography and literature.

《徐霞客游记》由明代地理学家、游记作家徐霞客创作，记录了他34年的中国之旅，并对地理、水文、地质、植物等现象均进行了详细记录，在地理学和文学上都卓有建树。

Xu Xiake recorded karst landforms, mountain landforms, red beds landforms, flowing water landforms, volcanic landforms, **periglacial**[②] landforms and applied landforms in his travels, describing 102 different landforms, among which the karst landforms in southwest China were particularly detailed.

徐霞客记录在游记中的地貌类型有喀斯特地貌、山岳地貌、红层地貌、流水地貌、火山地貌、冰缘地貌和应用地貌，他描述过的地貌形态名称多达102种，其中以中国西南地区喀斯特地貌尤为详细。

Xu Xiake's Travels **detailed**[③] the underground water. He divided the underground water into cold spring, hot spring, thermal spring and boiling spring while he recorded the use of underground water resources① in various ways, such as bathing, treatment, food processing, and so on.

《徐霞客游记》详细记载了地下水。他把水温不同的地下水分为冷水泉、温泉、热水泉和沸泉，同时他还记载了利用地下水资源的种种方法，如沐浴、治病、食品加工等。

佳句点睛　Punchlines

1. Xu Xiake, a famous traveler and geographer in the Ming Dynasty, wrote the masterpiece *Xu Xiake's Travels* to **document**② every step of his travels in China.

明代著名旅行家、地理学家徐霞客，著有《徐霞客游记》，记录了他游历中国的详细旅程。

2. Xu Xiake's writing falls under the ancient Chinese literary category of travel record literature, which used narrative and prose styles of writing to **portray**③ a person's travel experiences.

徐霞客的作品属于中国古代游记文学的范畴，以叙事性和散文性的写作方式来描述个人的旅行经历。

3. From the snowy passes of Sichuan, to the semi-tropical jungles of Guangxi, and to the mountains of Tibet, Xu Xiake wrote all his experiences and provided enormous amounts of details from his observations.

从四川的雪道,到广西的亚热带丛林,再到西藏的山脉,徐霞客写下了他所有的经历,并通过观察提供了大量细节。

情景对话 Situational Dialogue

A: Is it the China Tourism Day today?

B: Yes, it is. Since 2011, the Chinese travel authorities have designated May 19 each year as China Tourism Day in order to **commemorate**① Xu Xiake, the great Chinese traveler, writer and geographer.

A: Is he the author of *Xu Xiake's Travels?*

B: Yes. the first diary was recorded by Xu Xiake on May 19, 1613. Xu Xiake embarked on his first journey when he was 22 and then spent more than 30 years traveling all around China.

A: It was often commented that Xu Xiake's mother encouraged him to leave home and explore the outside world.

B: Because he loved to read about history, geography, exploration and traveling when he was very young.

A: Xu Xiake was also **deemed**② the first person in the world to study karst geomorphology.

B: Yes. His great courage, curiosity and **persistence**③ led him to numerous places that were unknown or even existed only in legends.

A: And his **meticulous**④ style also earned him the reputation of being a great writer.

B: No matter how tired he was or how poor the conditions of his lodging for the night, Xu Xiake would unfailingly write down his experiences, observations and discoveries in his diary every day.

A: 今天是中国旅游日吗?

B: 是的。自2011年起,中国旅游部门将每年的5月19日定为中国旅游日,以纪念伟大的中国旅行家、作家和地理学家徐霞客。

A: 他是《徐霞客游记》的作者吗?

B: 是的,徐霞客在1613年的5月19日写下了他的第一篇日记。他22岁时开始第一次旅行,然后花了三十多年的时间环游中国。

A: 经常有人说是徐霞客的妈妈鼓励他离家去探索外面的世界。

B: 因为他在很小的时候就喜欢读历史、地理、探险和旅游方面的书。

A: 徐霞客还被认为是世界上第一个研究喀斯特地貌的人。

B: 是的。他的勇气、好奇心和毅力驱使他去了许多未知的、甚至只存在于传说中的地方。

A: 徐霞客一丝不苟的风格也为他赢得了伟大作家的声誉。

B: 不管有多累,不管住宿条件有多差,他每天都坚持把自己的经历、观察和发现写在日记里。

生词注解 Notes

① hydrology /haɪˈdrɒlədʒɪ/ *n.* 水文学

② periglacial /ˌperɪˈɡleɪʃl/ *adj.* 冰缘的;冰边的

③ detail /ˈdiːteɪl/ vt. 详细叙述；详细列举

④ resource /rɪˈsɔːs/ n. 资源；智谋

⑤ document /ˈdɒkjʊm(ə)nt/ vt. 记录；证明

⑥ portray /pɔːˈtreɪ/ vt. 描绘；扮演

⑦ commemorate /kəˈmeməreɪt/ vt. 庆祝；纪念

⑧ deem /diːm/ vt. 认为；相信

⑨ persistence /pəˈsɪstəns/ n. 坚持不懈；毅力

⑩ meticulous /məˈtɪkjələs/ adj. 一丝不苟的；小心翼翼的

第二部分 现代科技成就

Part II Modern Achievements in Science and Technology

超级杂交水稻

Super Hybrid Rice

 导入语 Lead-in

水稻是主要的粮食作物之一,世界上半数以上的人口都以稻米为主食。超级杂交水稻,简称"超级稻",是指中国农业农村部根据《超级稻品种确认办法》确定的高产量超级稻品种,以中国工程院院士、"杂交水稻之父"袁隆平为代表的中国科学家通过杂交等科学方法,培育出一系列具有抗倒伏、抗病虫害、高产优质、"中大穗、粒多、粒重"等优良性状的水稻品种。这些超级杂交水稻为确保中国乃至世界的粮食安全和人民的幸福生活起到了极其重要的支撑作用。

文化剪影 Cultural Outline

China is the first country in the world to develop and **cultivate**① the **hybrid**② rice varieties successfully. In 1996, the Ministry of Agriculture proposed the Super Hybrid Rice Cultivation Plan in order to meet the demand for food. The plan was hosted by Yuan Longping, "Father of Hybrid Rice". And **heterosis**③ and three-line hybrids (**sterile**④ lines, inbred lines, and fertility restorer lines) were employed to achieve higher yields per mu.

中国是世界上第一个成功研发和推广杂交水稻的国家。农业部于1996年提出中国超级杂交水稻培育计划，以满足我国日益增长的粮食需求目标。该计划由"杂交水稻之父"袁隆平主持，利用杂种优势，选择优良亲本，用三系法（不育系、自交系、恢复系）取得更高的亩产和更佳的种植特性。

Since the **implementation**⑤ of Super Hybrid Rice Cultivation Plan, hybrid rice varieties have been cultivated in some major grain areas such as Hunan, Anhui, Sichuan, Henan, and Guangxi Provinces. In 2014, the representative super hybrid rice variety "Y Liangyou 900" in the fourth phase set a new high-yield record of 1026.7 kg per mu for 100 mu of contiguous plots, twice the average rice yield in China, which means super hybrid rice can be grown on a large scale over a large area.

超级杂交水稻培育计划自实施以来,已在湖南、安徽、四川、河南、广西等粮食主产区稳步推进。2014年,第四期超级杂交水稻代表品种"Y两优900"创造了百亩连片平均亩产1026.7千克的高产新纪录,两倍于中国水稻的平均产量,这意味着超级杂交水稻可以在大范围内实现大面积、大规模种植。

Compared with the **conventional**⑥ rice, super hybrid rice has dual advantages in breed and technology. In breed, it has the triple advantages of "medium and large spikes, multiple grains per plant, and grain weight per **panicle**⑦". In technology, the method of "combining morphological improvement and heterosis" is the core of super hybrid rice breeding, and it has achieved the goals of breeding in different periods.

跟常规水稻相比,超级杂交水稻具有品种和技术上的双重优势。在品种上,具有"中大穗、粒多、粒重"三重优势。在技术上,"形态改良与杂种优势利用相结合"的培育方式是中国超级杂交水稻育种的灵魂思想,实现了不同时期超级杂交水稻育种的目标。

佳句点睛　Punchlines

1. China is the first country in the world to successfully develop and popularize hybrid rice.
中国是世界上第一个成功研发和推广杂交水稻的国家。

2. Yuan Longping, "Father of Hybrid Rice", achieved a historic

breakthrough in hybrid rice with his research group in the 1970s.

"杂交水稻之父"袁隆平及其科研团队于20世纪70年代实现了杂交水稻的历史性突破。

3. Super hybrid rice has not only solved China's food crisis, but also helped more countries out of hunger and poverty.

超级杂交水稻不仅解决了中国的粮食危机问题,而且帮助更多的国家走出了饥饿与贫困的境遇。

情景对话 Situational Dialogue

A: Can I ask you some questions about super hybrid rice?

B: Yes, you can.

A: Why does super hybrid rice have such excellent **phenotype**[®]?

B: The super hybrid rice can achieve heterosis through hybridization. To put it clear, it is to combine the advantages of different strains of rice and further develop them.

A: Do planting conditions have a big impact on the yield of super hybrid rice?

B: In order to achieve high yield of rice, good varieties, fertile land, good methods are **indispensable**[®]. In China, more than 50% of the cultivated land in China are the low-yield fields. In recent years, our government has invested huge amounts of funds to transform the fields and carry out farmland water conservancy construction, which has laid the foundation for high yields. Good methods refer to proper farming methods, such as water and fertilizer management, pest control, and so on, which also have a great impact on yield.

A: Why the grains grown by super hybrid rice cannot be used directly for seed planting?

B: Genetically, because of the heterozygote, the offspring of super hybrid rice will have character **segregation**①. That is to say, the seeds of super hybrid rice cannot be planted again, and its excellent characteristics cannot be maintained. This is also one of the disadvantages of the super hybrid rice.

A: Thank you. Your explanation is simple but profound, making more sense to me on the super hybrid rice.

A: 我能问您一些关于超级杂交水稻的问题吗?

B: 可以。

A: 超级杂交水稻为什么能有如此优良的性状?

B: 通过杂交,超级稻可以实现杂种优势。通俗一点说,就是通过不同品系水稻之间的杂交,把它们的优点结合起来,并进一步发扬光大。

A: 种植条件对超级稻产量的影响大吗?

B: 为了实现水稻高产,良种、良田、良法缺一不可。尽管国内有百分之五十以上的耕地是中低产田,但近年来国家投入巨额资金改造田地、进行农田水利建设,为实现高产奠定了基础。良法指正确的耕作方式,比如水肥管理、病虫害防治等,对产量也有很大的影响。

A: 为什么超级稻种出的米粒不能直接用于继续种植呢?

B: 从遗传上来讲,超级稻是杂合子,后代会出现性状分离。也就是说,超级稻的种子再次种植,不能保持超级稻的优良性状。这也是目前超级稻的缺点之一。

A: 谢谢您。您的讲解深入浅出,使我对超级水稻有了更多的了解。

生词注解 Notes

① cultivate /ˈkʌltɪveɪt/ vt. 培育;种植

② hybrid /ˈhaɪbrɪd/ adj. 杂种的;混合的

③ heterosis /hetəˈrəʊsɪs/ n. 杂种优势

④ sterile /ˈsterəl/ adj. 不育的;不孕的

⑤ implementation /ˌɪmplɪmenˈteɪʃn/ n. 实施;执行

⑥ conventional /kənˈvenʃnl/ adj. 传统的;常见的

⑦ panicle /ˈpænɪkl/ n. 圆锥花序

⑧ phenotype /ˈfiːnətaɪp/ n. 表型;显型

⑨ indispensable /ˌɪndɪˈspensəbl/ adj. 不可或缺的

⑩ segregation /ˌsegrɪˈgeɪʃn/ n. 隔离;分离

人工合成结晶牛胰岛素

Synthetic Crystalline Bovine Insulin

导入语 Lead-in

胰岛素是一种蛋白质类激素，主要用于控制血糖。1921年，加拿大医师班廷发现了胰岛素并将其应用于临床，挽救了过去不治的糖尿病患者，开创了临床医学应用胰岛素的先河。但是，直到20世纪50年代，人工合成蛋白质依然是科学领域无人攀登的高峰。1965年9月，新中国的科技先驱在经济基础羸弱、西方发达国家严密封锁的艰难环境下刻苦攻关，首次获得了具有生物活性的人工合成结晶牛胰岛素。这是世界上第一个在体外全合成、仍具有完整结构的功能性蛋白质，开创了人工合成蛋白质的新纪元。

文化剪影 Cultural Outline

The Chinese scientists first **synthesized**① crystalline bovine **insulin**② in September, 1965. Members in the research group from the Chemistry Department of Beijing University, Shanghai Institute of Biochemistry, CAS and Shanghai Institute of Organic Chemistry, CAS started to study the synthetic crystalline bovine insulin in 1958. After six-year **collaborative**③ efforts, they obtained successfully the crystals of total synthesis of bovine insulin, which was the first functional crystalline protein being fully synthesized in the world. It is a crucial step toward understanding and exploring the secrets of life.

1965年9月，中国科学家首次实现了人工合成结晶牛胰岛素。从1958年开始，由北京大学化学系、中国科学院上海生物化学研究所、中国科学院上海有机化学研究所的研究者组成的团队经过6年多的努力，于1965年9月成功获得了人工全合成的牛胰岛素的晶体，这是世界上第一个人工合成的具有生物活性的结晶蛋白质。它标志着人类在认识生命、探索生命奥秘的征途中迈出了具有里程碑意义的重要一步。

Synthetic crystalline bovine insulin, fully synthesized in vitro with a complete structure, is the functional protein. Its structure, biological **vitality**④, physical and chemical properties, and crystalline shape are exactly the same as those of natural bovine insulin. It has made a significant **impact**⑥ on life sciences.

人工合成结晶牛胰岛素在体外全合成且仍具有完整结构的功能性蛋白质。其结构、生物活力、物理化学性质、结晶形状都和天然的牛胰岛素完全一致,在生命科学发展史上产生了重大影响。

Insulin, a protein hormone **secreted**⑥ by the **pancreas**⑦ of animals, regulates blood glucose concentration. So far, it has been used as an effective medicine for treating diabetes clinically. Synthetic crystalline bovine insulin reduces the cost of diabetes treatment and brings better treatment results.

胰岛素是一种由动物胰脏分泌的、用于调节血糖浓度的蛋白质激素,至今在临床上一直作为治疗糖尿病的特效药物。人工合成结晶牛胰岛素降低了糖尿病的治疗成本,同时也带来了更好的治疗效果。

佳句点睛　Punchlines

1. The Chinese scientists first synthesized crystalline bovine insulin in September, 1965.

1965年9月,中国科学家首次获得了人工合成结晶牛胰岛素。

2. The successful obtainment of synthetic crystalline bovine insulin is a crucial step toward understanding and exploring the secrets of life.

人工合成结晶牛胰岛素的成功研制标志着人类在认识生命、探索生命奥秘的征途中迈出了关键一步。

3. Synthetic crystalline bovine insulin reduces the cost of **diabetes**[®] treatment, and brings better efficacy.

人工合成结晶牛胰岛素降低了糖尿病的治疗成本,带来了更好的治疗效果。

情景对话 Situational Dialogue

A: Do you know insulin?

B: Yes, of course.

A: You are not a medical major. How do you know about it?

B: My mother got diabetes ten years ago. She has been controlling her blood glucose level by **injecting**[®] insulin.

A: Do you know that China is the first country obtaining synthetic crystalline bovine insulin?

B: I have no idea about it. Can you tell me something about it?

A: Insulin, a protein hormone secreted by the pancreas of animals, regulates blood glucose concentration. In 1921, insulin was discovered by Canadian physician G. Banting and used clinically. In 1955, its structure was **clarified**[®] by British Scientist F. Sanger, who won the Nobel Prize in Chemistry. From the late 1950s, Chinese scientists have begun to study the artificial synthesized living matter. After six years of hard work, a small bottle of white crystals was obtained in September, 1965, which was the first synthesis crystalline bovine insulin in the world.

B: The scientists of our country are too great because our economic foundation at that time was weak and there was no favorable scientific research environment.

A: I agree with you. So let's work harder now.

A: 你知道胰岛素吗?

B: 当然知道。

A: 你不是医学专业的学生,怎么会了解胰岛素?

B: 我母亲得糖尿病已经十多年了,一直靠注射胰岛素来控制血糖。

A: 那你知道中国是第一个实现人工合成结晶牛胰岛素的国家吗?

B: 这我还不了解,你能给我讲讲吗?

A: 胰岛素是一种由动物胰脏分泌、可以调节血糖水平的蛋白质激素。加拿大医师班廷于1921年发现胰岛素并将其用于临床。英国科学家桑格于1955年阐明了它的化学结构,并因此获得诺贝尔化学奖。我国科学家从20世纪50年代末开始探索人工合成生命物质,经过六年多的努力,在1965年9月制得一小瓶雪白的结晶体,这是世界上首次获得的具有生物活性的人工合成结晶牛胰岛素。

B: 我国的科学家太伟大了,那时候正是我国经济基础薄弱时期,没有现在这么好的科研条件。

A: 我同意你的看法。所以,我们现在更要努力学习了。

生词注解 Notes

① synthesize /ˈsɪnθəsaɪz/ *vt.* 合成;综合

② insulin /ˈɪnsjəlɪn/ *n.* 胰岛素

③ collaborative /kəˈlæbərətɪv/ *adj.* 合作的;协作的

④ vitality /vaɪˈtælətɪ/ *n.* 活力;生命力

⑤ impact /ˈɪmpækt/ *n.* 影响;效果

⑥ secrete /sɪˈkriːt/ *vt.* 分泌

⑦ pancreas /ˈpæŋkrɪəs/ *n.* 胰腺

⑧ diabetes /ˌdaɪəˈbiːtiːz/ *n.* 糖尿病

⑨ inject /ɪnˈdʒekt/ *vt.* 注射;注入

⑩ clarify /ˈklærəfaɪ/ *vt.* 澄清;阐明

现代科技成就

"两弹一星"

"Two Bombs and One Satellite"

 导入语　Lead-in

"两弹一星"指核弹、导弹和人造卫星，是新中国引以为豪的伟业之一。新中国成立后，面对严峻的国内外形势，党中央毅然做出了发展中国核事业和航天事业的伟大战略决策，大批出类拔萃的科技工作者克服了艰苦的环境和薄弱的经济与技术基础带来的困难，成功突破了核弹、导弹和人造卫星等尖端技术。以"两弹一星"为核心的国防尖端科技的辉煌成就，不仅是中国国防现代化的伟大成就，也是中国现代科学技术事业发展的重要标志。"两弹一星"带动了中国现代科学技术的发展，填补了许多学科空白，为中国实现技术发展的跨越积累了宝贵经验。

科技成就

文化剪影　Cultural Outline

From the 1950s to the 1970s, in an environment of weak economic and technological strength and hardship, with the full cooperation and strong support of all walks of life across the country, and with less investment and in a relatively short period of time, China succeeded in its own research and development of "Two Bombs and One Satellite", i.e., the atomic bomb, the missile and the man-made earth satellite, have created a **miracle**[①] on the history of modern Chinese science and technology.

20世纪50至70年代,在经济条件落后、技术力量薄弱的艰苦环境下,在全国各行各业通力协作和大力支持下,中国用较少的投入和较短的时间,成功独立自主研发了"两弹一星",即原子弹、导弹和人造地球卫星,创造了中国现代科学技术发展史上的奇迹。

China tested its first atomic bomb in 1964 and **hydrogen**[②] bomb in 1967, and successfully launched its first satellite (DFH-1) in 1970. The success of "Two Bombs and One Satellite" Project is a glorious cause created by the Chinese nation in the second half of the 20th century. In the course of research and development, the achievements made by scientific researchers in missiles, satellites, remote sensing and control have also laid the foundation for the further development of China's **aerospace**[③] industry.

中国在1964年成功爆炸第一颗原子弹,在1967年成功爆炸第一枚氢弹,在1970年成功发射了第一颗人造卫星(东方红一号)。"两弹一星"工程的成功是20世纪下半叶中华民族创建的辉煌伟业。在研发过程中,科研人员在导弹、人造卫星、遥感与制控等方面取得的成果也为中国航天事业的进一步发展打下了基础。

During the research of the early nuclear and space project, China has formed the spirit of "Two Bombs and One Satellite". This spirit is rich in its **connotation**④, mainly involving love for the motherland, **dedication**⑤ to the career, self-dependence, hard-working, collocation and innovation. This spirit has been and will be carried forward in the development of space industry in China. And it is the spiritual driving force for the great **rejuvenation**⑥ of Chinese nation.

在"两弹一星"事业研发的过程中,逐步形成了热爱祖国、无私奉献、自力更生、艰苦奋斗、大力协同、勇于登攀的"两弹一星"精神。这种精神在中国航天事业发展中持续得到了弘扬,是中华民族伟大复兴的精神动力。

佳句点睛 Punchlines

1. "Two Bombs and One Satellite" refer to the atomic bomb, the missile and the man-made earth satellite.
"两弹一星"是指核弹、导弹和人造卫星。

2. "Two Bombs and One Satellite", which was successfully devel-

oped in the 1960s, **demonstrated**① China's national strength and technological capabilities to the world.

20世纪60年代成功研制的"两弹一星"向世界展示了中国的国力和技术能力。

3. The spirit of "Two Bombs and One Satellite" is the spiritual driving force for the great rejuvenation of Chinese nation.

"两弹一星"精神是中华民族伟大复兴的精神动力。

 情景对话　Situational Dialogue

A: Today I visited "Two Bombs and One Satellite" Memorial Museum in the University of the Chinese Academy of Sciences.

B: Really? Can you tell me about it?

A: Very impressive! I'm deeply impressed by their scientific spirit of the older generation, especially the spirit of "Two Bombs and One Satellite".

B: I know our government has awarded "Two Bombs and One Satellite Merit Award" to the scientists who have made outstanding **contributions**② in the study of "Two Bombs and One Satellite".

A: Yes, on September 18, 1999, on the eve of the 50th anniversary of the founding of the People's Republic of China, twenty-three scientists involved in the project were awarded.

B: I know Qian Xuesen, "Father of Chinese Missiles", was honored as one of the "People Who Moved China" in 2007. I was moved to tears when I watched the Awards Party.

A: Yes, I was moved again and again in the museum. They have overcome various difficulties and **obstacles**⑥ and made outstanding contributions to the development of space industry in China.

B: I'll definitely visit this museum when I'm free. I want to learn from these great scientists and carry forward the spirit of "Two Bombs and One Satellite".

A: Yes, let's work hard together.

A: 今天我去中国科学院参观了"两弹一星"纪念馆。

B: 是吗？能给我讲讲吗？

A: 很震撼！我深深折服于老一辈科学家的科学精神，还有"两弹一星"精神。

B: 我知道我国曾为参与研制"两弹一星"并做出突出贡献的科学家授予了"两弹一星功勋奖章"。

A: 是的，1999年9月18日，在庆祝中华人民共和国成立50周年之际，我国对23位在"两弹一星"研制过程中做出突出贡献的科学家进行了表彰。

B: 我知道"中国导弹之父"钱学森，他在2007年被评为"感动中国"年度人物。我当时看颁奖晚会时都感动得哭了。

A: 是的，我在博物馆也一次次被感动，科学家们克服各种困难，为我国航天事业的发展做出了突出贡献。

B: 有时间我也一定去参观一次，向这些伟大的科学家学习，去

科技成就

发扬"两弹一星"精神。

A：是的，让我们一起加油。

生词注解 Notes

① miracle /ˈmɪrəkl/   n. 奇迹；奇迹般的人或物

② hydrogen /ˈhaɪdrədʒən/   n. 氢

③ aerospace /ˈeərəʊspeɪs/   n. 航空航天；宇航

④ connotation /ˌkɒnəˈteɪʃn/   n. 内涵；隐含意义

⑤ dedication /ˌdedɪˈkeɪʃn/   n. 奉献；献身

⑥ rejuvenation /rɪˌdʒuːvəˈneɪʃn/   n. 复兴；返老还童

⑦ demonstrate /ˈdemənstreɪt/   vt. 展示；证明

⑧ contribution /ˌkɒntrɪˈbjuːʃn/   n. 贡献；捐献

⑨ obstacle /ˈɒbstəkl/   n. 障碍；障碍物

汉字激光照排技术

The Laser Phototypesetting[①] System of Chinese Characters

 导入语　Lead-in

汉字激光照排系统是20世纪80年代由北京大学王选教授带领科研团队研制的一项伟大发明，被誉为"汉字印刷术的第二次发明"。这项发明针对汉字字数多、存储量大的特点和难点，研制了高分辨率字形的高倍率信息压缩技术和高速复原方法，率先设计出相应的专业芯片，在世界上首次使用了控制信息参数描述笔画特性的方法，并取得一项欧洲专利和八项中国发明专利。

汉字激光照排系统的发明和产业化应用，替代了中国沿用上百年的铅字印刷，彻底改变了印刷行业的命运，推动了中国报业和印刷出版

业的技术革命,使中国印刷业告别了"铅与火"的时代,跨入了"光与电"的时代。

文化剪影　Cultural Outline

The laser phototypesetting system of Chinese characters consists of three parts such as input **device**②, computer information processing, and laser scanning recording. Input device can be the paper tape or the diskette, or the communication system. Information processing system is composed of an operation **console**③, an electronic computer, and a hard disk drive. Following the input codes and operation control instructions, this system completes four main programs of control, layout, arrangement and exposure. Meanwhile, it controls, directs, **dispatches**④ and monitors the printer. In the part of laser scanning and recording, the dot-matrix glyph information output processed by the computer is recorded by the laser plane line scanning host.

汉字激光照排系统由输入设备、电子计算机信息处理和激光扫描记录三部分组成。输入部分可以使用纸带或软磁盘等设施,也可接受通信系统的输入。电子计算机信息处理部分由操作控制台、电子计算机和硬磁盘驱动器组成,按照输入代码和操作控制指令,完成控制、编排、拼排和曝光这四个主要程序,并对整机起着控制、指挥、调度和监视的作用。激光扫描记录部分由激光平面线扫描主机,记录经计算机处理后输出的点阵字形信息。

The laser phototypesetting technology of Chinese characters can compress Chinese character information with high power, reduce characters with high speed and keep the original appearance of characters. With its advantages of high efficiency, flexible layout and complete character library, it has completely transformed the traditional Chinese publishing and printing industry that has lasted for hundreds of years.

汉字激光照排技术具有高倍率汉字信息压缩技术、高速度还原技术和不失真的文字变倍技术等特点,以效率高、版面灵活和字库齐全等优势彻底改变了延续上百年的中国传统出版印刷业。

The laser phototypesetting system of Chinese characters **eliminates**[5] the lead pollution, reduces energy **consumption**[6], shortens the publishing cycle, increases the publishing varieties, and improves the printing quality. This digital **transformation**[7] allows Chinese civilization and printing technology to keep up with the pace of the informative world.

汉字激光照排系统消除了铅毒污染,降低了能耗,缩短了出版周期,增加了出版品种,提高了印刷质量。这种数字化的转变让汉字文明和印刷技术跟上了世界信息化和网络化的步伐。

佳句点睛　Punchlines

1. The laser phototypesetting system of Chinese characters was a great invention developed in the 1980s by Professor Wang Xuan and

his scientific research team.

汉字激光照排系统是20世纪80年代由王选教授带领科研团队攻坚克难研制出来的伟大发明。

2. The laser phototypesetting technology of Chinese characters has the characteristics of high-rate Chinese character information compression technology, high-speed restoration technology and distortion-free text-to-text technology.

汉字激光照排系统具有高倍率汉字信息压缩技术、高速度还原技术和不失真的文字变倍技术等特点。

3. The laser phototypesetting system of Chinese characters has achieved "the second revolution" of printing technology in China.

汉字激光照排系统实现了中国印刷技术的"第二次革命"。

情景对话 Situational Dialogue

A: Congratulations! You've got such high points in National College Entrance Examination. Which university do you want to go to? What major do you want to choose?

B: I want to learn computational mathematics in Peking University.

A: Very good idea. But you have to know that this discipline involves knowledge of mathematics, physics, computer science, operations research and control science, and so on, which is somewhat complicated[®].

B: I know. A few days ago, I read an article about Wang Xuan, who developed the laser phototypesetting system of Chinese characters. It was computational mathematics that he learned in Peking University when he was young.

A: Yes. It was in 1950s when he was in the university. At that time, **computational**[®] mathematics was a new discipline and the information technology was also quite backward in China. But Wang Xuan, with his research team, began to study information processing technology of Chinese characters in the mid-1970s based on what he had learned. After more than ten years of hard work, he developed the laser phototypesetting technology of Chinese characters, which is known as "the second revolution" of printing technology in China.

B: Yes, printing is one of the Four Great Inventions in ancient China. But later, this technique has fallen behind the western countries. It is the laser phototypesetting technology of Chinese characters developed by Professor Wang Xuan and his team that has boosted the printing industry in China.

A: Now the calculation problem is common in all fields of modern society. And a sea of data in various industries need to be calculated and analyzed to find out some rules and laws. I hope you can become a person with outstanding achievements like Professor Wang Xuan someday.

B: Thank you for your encouragement. I'll work harder.

科技成就

A：恭喜你，高考成绩不错。你想报考哪个学校？学什么专业？

B：我想报考北京大学的计算数学专业。

A：非常好的想法。但你要知道，这个学科涉及数学、物理学、计算机科学、运筹学和控制科学等多个学科的知识，是有一定难度的。

B：我知道。前几天我看了一篇关于汉字激光照排技术的发明者——北京大学王选教授的文章，他当时学的就是计算数学。

A：是的，王选教授上大学时应该是20世纪50年代，那时候我国信息技术相当落后，计算数学是新兴学科，但王选用他所学的知识，同他的科研团队从20世纪70年代中期开始研发汉字信息处理，经过十多年的努力，研发了汉字激光照排技术，被誉为"汉字印刷术的第二次发明"。

B：是的，印刷术是我国古代四大发明之一，但后来我国印刷术的发展就落后于西方发达国家了。是王选教授及其团队研发的汉字激光照排技术，实现了我国印刷业产业化的发展。

A：计算问题是现代社会各个领域普遍存在的共同问题。各个行业都有大量数据需要计算，都可以通过数据分析来掌握事物发展的规律。我希望你也能像王选教授那样，成为一个有突出贡献的人。

B：谢谢您的鼓励！我会更努力的。

生词注解 Notes

① phototypeset /ˈfəʊtəˈtaɪpset/ vt. 照相排版

② device /dɪˈvaɪs/ n. 设备；装置

③ console /kənˈsəʊl/ n. 控制台；操纵台

④ dispatch /dɪˈspætʃ/ vt. 派遣；发送

⑤ eliminate /ɪˈlɪmɪneɪt/ vt. 消除；排除

⑥ consumption /kənˈsʌmpʃn/ n. 消费；消耗

⑦ transformation /ˌtrænsfəˈmeɪʃn/ n. 转化；转换

⑧ complicated /ˈkɒmplɪkeɪtɪd/ adj. 复杂的；难懂的

⑨ computational /ˌkɒmpjuˈteɪʃənl/ adj. 使用计算机的；计算的

科技成就

中国航天工程

China Aerospace Industry

导入语　Lead-in

中国航天工程始建于1956年,经历了艰苦创业、稳定发展、改革振兴和走向世界等重要时期,迄今已经形成了配套完整的研究、设计、生产和试验体系,建立了能发射各类卫星和载人飞船的航天器发射中心和由国内各地面站、远程跟踪测量船组成的测控网、多种卫星应用系统和具有一定水平的空间科学研究系统,培育了一支素质好、技术水平高的航天科技队伍。中国航天发展的一次次突破创造了中国航天史和世界航天史上的新奇迹,承载了中华民族的光荣与梦想,彰显了中国的繁荣与富强。

文化剪影　Cultural Outline

The aerospace industry includes three parts such as space technology, space application, and space science. It is a highly **integrated**① modern science and technology and the latest achievements of basic industry while it is an important symbol of a country's scientific and technological level and comprehensive national strength. Since its establishment in 1956, China's aerospace industry has independently developed and launched communication satellites, remote sensing satellites, **meteorological**② satellites, and navigation satellites, which have been widely used in many fields of national economy and social life. The Long-March series of carrier rocket have promoted the development of the satellites and their applications as well as manned space technology. Shenzhou manned spacecrafts have achieved some major breakthroughs, such as our participation in space test activities and our successful implementation of space **evacuation**③ activities. And Chang'e project has demonstrated our ability of lunar exploration.

航天工程包括空间技术、空间应用和空间科学三部分，是现代科学技术和基础工业最新成绩的高度综合，也是一个国家科学技术水平和综合国力的重要标志。自1956年中国航天事业创建以来，自主研制发射的通讯卫星、遥感卫星、气象卫星、导航卫星，已经广泛应用于国民经济和社会生活的诸多领域。长征系列运载火箭推动了中国卫星及其应用以及载人航天技术的发展。神州载人飞船在参与空间试验活动、成功实施空间出舱活动等方面取得了重大突破。嫦娥工程彰显了中国的深空探测能力。

China aerospace industry has made a series of brilliant achievements represented by "Two Bombs and One Satellite", manned spaceflight and lunar exploration. On April 24, 1970, the first man-made satellite "DFH-1" was launched successfully, **symbolizing**④ China became the fifth country to develop and launch satellites independently in the world. On November 26, 1975, the first recoverable satellite was successfully launched, and China created a miracle in the history of world aerospace. On October 15, 2003, "Shenzhou V" spacecraft entered space, marking **fulfilling**⑤ China's dream of travelling in the space. On October 24, 2007, the successful launch of "Chang'e-1" marked a new stage in China's aerospace technology from low-altitude exploration to deep-space exploration.

中国航天事业取得了以"两弹一星"、载人航天工程和探月工程为代表的一系列辉煌成就。1970年4月24日,我国第一颗人造卫星"东方红一号"进入太空,中国成为世界上第五个能独立研制、发射人造地球卫星的国家。1975年11月26日,首发返回式卫星成功,中国创造了世界航天史上的一个奇迹。2003年10月15日,"神州五号"飞船进入太空,标志着华夏飞天圆梦。2007年10月24日,"嫦娥一号"的成功发射,标志着中国航天技术从低空探索迈向深空探索的新阶段。

The space vehicles of "Long-March" "Shenzhou" "Chang'e" and "Tiangong" have witnessed the great achievements in China aerospace industry, marking a new level of China space exploration. And now, China has formed a complete system in research, design, production and test in space industry. She has achieved rapid development in space technology, space application and space science and gradually become a major aerospace power.

"长征""神州""嫦娥"和"天宫"等航天飞行器见证了中国航天的伟大成就,标志着中国太空探索的新高度。目前,中国航天已经形成了完整配套的研究、设计、生产和试验体系,在空间技术、空间应用和空间科学三大领域取得了深远的发展,逐渐成为一个航天大国。

佳句点睛 Punchlines

1. The aerospace industry involves three parts such as space technology, space applications and space science.
航天工程包括空间技术、空间应用和空间科学三部分。

2. The space vehicles of "Long-March" "Shenzhou" "Chang'e" and "Tiangong" have witnessed the great achievements in China aerospace industry, marking a new level of China space exploration.
"长征""神州""嫦娥"和"天宫"等航天飞行器见证了中国航天的伟大成就,标志着中国太空探索的新高度。

3. China successfully launched its first satellite "DFH-1" on April 24, 1970.

1970年4月24日,中国成功发射了第一颗人造卫星"东方红一号"。

情景对话 Situational Dialogue

A: Good morning, boys and girls. In today's lecture on aerospace, we'll learn something about the manned space flight in China. Who do you guys know about the **astronaut**①?

B: Yang Liwei, who is the first astronaut in China to fly to space.

A: Good. Who else?

C: Jing Haipeng, who has been sent into the space for three times.

D: And Liu Yang and Wang Yaping, who are female astronauts flying into space in China.

A: Very good. You know the astronauts very well. Do you know who proposed the plan of manned spaceflight in China?

All: I don't know.

A: Qian Xuesen, a great scientist, who is the founder of China Manned Spaceflight. The research on manned spaceflight can be traced back to the 1970s. After the first satellite "DFH-1" was launched successfully, Qian Xuesen proposed that China develop the manned spaceflight program — "Project 714". But at that time China's comprehensive national strength and industrial foundation were very weak, and this project was put on hold. It was not until September 1992 when the central government decided to **implement**⑦ the manned space-flight program and determined the "three-step" strategy. After years of hard work of numerous scientists, you can see the achievements our country has made in the aerospace industry.

A: 早上好,同学们,在今天的航天知识讲座中,我们将了解中国载人航天的发展。你们知道的航天员都有谁?

B: 杨利伟,他是我国第一位飞上太空的航天员。

A: 很好。还有吗?

C: 景海鹏,他曾经三次飞上太空。

D: 还有刘洋和王亚平,她们是飞上太空的女宇航员。

A: 非常好。看来同学们对宇航员还是非常了解的。那你们知道我国的载人航天发展计划是谁提出来的吗?

All: 不知道。

A: 是钱学森,一位伟大的科学家,他是我国载人航天事业的奠基人。我国的载人航天研究可以追溯到20世纪70年代,在我国第一颗人造地球卫星"东方红一号"上天后,钱学森就提出中国要搞载人航天,国家当时将这个项目命名为"714工程"。但是,那时候我国的综合国力和工业基础都非常薄弱,这个项目就被搁置了。一直到1992年9月,中央决策实施载人航天工程并确定了"三步走"的发展战略。经过无数科研工作者多年的努力,你们才看到了我国航天事业取得的成绩。

生词注解 Notes

① integrated /ˈɪntɪgreɪtɪd/ *adj.* 综合的；互相协调的

② meteorological /ˌmiːtɪərəˈlɒdʒɪkl/ *adj.* 气象的；气象学的

③ evacuation /ɪˌvækjuˈeɪʃn/ *n.* 疏散；撤离

④ symbolize /ˈsɪmbəlaɪz/ *vt.* 象征；用符号表现

⑤ fulfil /fʊlˈfɪl/ *vt.* 履行；完成

⑥ astronaut /ˈæstrənɔːt/ *n.* 宇航员；航天员

⑦ implement /ˈɪmplɪment/ *vt.* 实施；执行

南京长江大桥

Nanjing Yangtze River Bridge

导入语　Lead-in

　　南京长江大桥是第一座完全由中国工程师设计、由劳动人民建造的特大双层式铁路、公路两用桥。大桥建设历时8年,于1960年1月18日建成,是新中国战胜自然灾害和复杂地质条件、克服重重经济困难、勇于面对严峻的国内外形势进行的一次伟大尝试,有"争气桥"之称,在中国和世界桥梁史上具有重要意义。南京长江大桥的建造,不仅为南京提供了重要的社会发展资源,也促进了中国各区域之间的紧密联系,向世界展示了中国人民的智慧和力量。长江大桥是南京市的标志性建筑、江苏省的文化符号,也是"新金陵四十八景"之一。1996年以"世界最长的公路、铁路两用桥"被载入《吉尼斯世界纪录大全》。2016年入选《首批中国20世纪建筑遗产名录》。

文化剪影 Cultural Outline

The Nanjing Yangtze River Bridge, completed and open for traffic in 1960, is a double-decked road-rail **truss**① bridge across the Yangtze River between Pukou District and Gulou District in Nanjing. Its upper deck **spans**② 4,589 meters linking China National Highway 104. Its lower deck, with a double-track railway, is 6,772 meters long, and becomes one of the major North-South **transportation**③ systems by linking the Tianjin-Pukou Railway and the Shanghai-Nanjing Railway, which reduces the train crossing time from one and half hours by ferry to two minutes.

南京长江大桥于1960年建成通车,位于南京市浦口区和鼓楼区之间,是一座双层式公路、铁路两用桥梁。上层公路桥长4589米,连接104国道。下层为双轨复线铁路桥,全长6772米,连接津浦铁路与沪宁铁路,将火车过江时间由过去靠轮渡的一个半小时缩短为两分钟,成为贯通中国南北的交通大动脉之一。

The Nanjing Yangtze River Bridge was designed and built by our country. The completion and opening of the bridge ushered in a new era of China's "self-reliance" in the construction of large-scale bridges, indicating that China's bridge construction has reached the world advanced level, with great economic, political and strategic significance. Now, as a symbol of Nanjing, the Nanjing Yangtze River Bridge has become a cultural symbol of Jiangsu Province.

南京长江大桥是中国自行设计并建造而成的。它的建成通车开创了中国"自力更生"建设大型桥梁的新纪元，标志着中国的桥梁建设达到世界先进水平，具有巨大的经济意义、政治意义和战略意义。现在，作为南京的标志性建筑，南京长江大桥已经成为江苏省的文化符号。

The **components**[④] of the Nanjing Yangtze River Bridge, such as the upper-deck-road-truss bridge, the lower-deck-rail-truss bridge, the approach bridge, the **piers**[⑤], the **symmetrical**[⑥] bridge tower, the status on the bridge, and the **magnolia**[⑦] lampposts, make it not only a complete building spanning the Yangtze River, but also an architectural artwork with a heavy sense of history.

南京长江大桥包含公路桥、铁路桥、引桥、桥墩、桥头堡、铸铁浮雕和玉兰灯柱，其不仅是一座横跨长江的完整建筑，更是一件具有厚重历史感的建筑艺术品。

科技成就

 佳句点睛 Punchlines

1. The Nanjing Yangtze River Bridge, completed and open for traffic in 1960, is a double-decked road-rail truss bridge.

南京长江大桥于1960年建成通车,是一座双层式公路、铁路两用桥梁。

2. The Nanjing Yangtze River Bridge was the first heavy bridge designed and built in China.

南京长江大桥是中国自行设计和建设的第一座重型桥梁。

3. The completion and opening of the Nanjing Yangtze River Bridge made it one of the major north-and-south transportation arteries in China.

南京长江大桥的建成通车使其成为贯通中国南北的交通大动脉之一。

 情景对话 Situational Dialogue

A: Wow! I finally see the Nanjing Yangtze River Bridge I've been dreaming of.

B: Gorgeous! Where do we go to at first?

A: Let's go to the Great Fort first. It's said that there's a **sculp-**

ture[⑧] of Chairman Mao in it.

B: Bravo!

A: The completion of the Yangtze River Bridge demonstrated the unity and strength of the Chinese people.

B: Yes, the Nanjing Yangtze River Bridge is the first double-decked road-rail truss bridge designed and constructed by our own expertise. Nowadays, Chinese bridge manufacturing is close to half of the global market share, and the Nanjing Yangtze River Bridge is one of the starting points of the Chinese bridge-building legend.

A: Here we are. The sculpture of Chairman Mao is so great! Let's take a picture with it.

B: Okay.

A: 哇,我终于看到了梦寐以求的南京长江大桥了!

B: 太震撼了!我们先看哪里?

A: 先去看大桥头堡吧,据说里面有毛主席雕塑。

B: 太棒了!

A: 长江大桥的建成通车体现了中国人民的团结和力量。

B: 是的,南京长江大桥是第一座中国自行设计、建造的双层式铁路、公路两用桥。如今中国桥梁制造已经接近全球市场份额的一半,而南京长江大桥正是中国造桥传奇的起点之一。

A: 到了,毛主席雕像真伟岸!我们来合个影吧。

B: 好的。

生词注解 Notes

① truss /trʌs/　n. 桁架

② span /spæn/　vt. 跨越；持续

③ transportation /ˌtrænspɔːˈteɪʃn/　n. 运输；运输工具

④ component /kəmˈpəʊnənt/　n. 组成部分；成分

⑤ pier /pɪə(r)/　n. 桥墩；直码头

⑥ symmetrical /sɪˈmetrɪkl/　adj. 匀称的；对称的

⑦ magnolia /mæɡˈnəʊliə/　n. 玉兰类植物；木兰

⑧ sculpture /ˈskʌlptʃə(r)/　n. 雕塑；雕刻

港珠澳大桥

Hong Kong-Zhuhai-Macao Bridge

导入语 Lead-in

　　港珠澳大桥跨越伶仃洋海域，东接香港特别行政区，西接广东省珠海市和澳门特别行政区，是一项超大型跨海交通工程。大桥总长约55公里，设计使用寿命120年。历经6年筹划和9年建设，港珠澳大桥于2018年10月24日开通运营。港珠澳大桥作为世界总体跨度最长、钢结构桥体最长、海底沉管隧道最长的跨海大桥，也是公路建设史上技术最复杂、施工难度最大、工程规模最庞大的大桥，体现了中国的综合国力和自主创新能力。这是一座圆梦桥、同心桥、自信桥和复兴桥。

 文化剪影　Cultural Outline

The Hong Kong-Zhuhai-Macao Bridge, linking Hong Kong to the east, and Zhuhai and Macao to the west, is a combination of three cable-stayed bridges, an undersea tunnel and four **artificial**[①] islands.

港珠澳大桥东连香港，西接珠海和澳门，由三座斜拉桥、一条海底隧道和四个人工岛组成。

The completion of the Hong Kong-Zhuhai-Macao Bridge is of great **significance**[②] to the construction and economic development of the Greater Bay Area, providing convenient, smooth, safe and **efficient**[③] services for the local people.

港珠澳大桥的建成对大湾区的建设和经济发展具有重要意义，为当地人们提供了便捷、畅通、安全和高效的服务。

It will help improve personnel and trade exchanges between Hong Kong, Guangdong and Macao, and enhance the **comprehensive**[④] competitiveness of the Greater Bay Area and the Pearl River Delta.

港珠澳大桥将促进香港、广东和澳门三地间的人员流动和贸易往来，并增强大湾区和珠江三角洲的综合竞争力。

现代科技成就

佳句点睛 Punchlines

1. I can't help but marvel at the **ingenuity**① and **incredible**② **fortitude**③ of Chinese experts and the engineering construction teams who have created the world miracle.

中国专家和工程团队创造了世界奇迹,他们的聪明才智和惊人毅力让我不禁惊叹。

2. The bridge helps **foster**④ economic integration of the Greater Bay Area while enabling the competitiveness and complementary function of the cities.

这座大桥有助于促进大湾区的经济一体化,同时增强城市间的竞争力和互补功能。

3. Five types of cross-boundary vehicle are allowed to travel on the bridge, including coaches, shuttle buses, passenger car and lorries.

五类跨境车辆可在桥上行驶,包括旅游车、穿梭巴士、小客车和货车。

情景对话 Situational Dialogue

A: Linda, what are you going to do this Sunday?

B: I plan to go to Hong Kong for shopping, so I'm reading the travelers' tips for crossing China's mega bridge.

A: Is it the Hong Kong-Zhuhai-Macao Bridge? It is said that it's the world's longest sea bridge and connects three regions.

B: Yes, the Hong Kong-Zhuhai-Macao Bridge spans the Pearl River Delta in southern China. With the bridge in place, travelling time between Zhuhai and Hong Kong would cut down from about four hours to 30 minutes.

A: Wow, it's really a **spectacular**① project.

B: What's more, the shuttle bus service runs 24 hours a day as frequent as every 5 minutes. Facilities offering cross-boundary services also opens 24 hours a day.

A: It's so convenient. Have a good day!

B: Thank you.

A: 琳达，这周日你打算做什么？

B: 我要去香港购物，所以我正在阅读跨越中国大桥的旅行小贴士。

A: 是港珠澳大桥吗？据说它是世界上最长的跨海大桥，连接着三个地区。

B: 是的，港珠澳大桥跨越了中国南部地区的珠江三角洲。有了这座大桥，从珠海到香港的时间将从约4个小时缩短到30分钟。

A: 哇，这真是个了不起的工程。

B: 而且，穿梭巴士每5分钟一班，全天候服务。提供跨境业务的设施也是24小时开放。

A: 简直太方便了,祝你玩得开心!
B: 谢谢你。

生词注解 Notes

① artificial /ˌɑːtɪˈfɪʃl/ *adj.* 人造的;人工的

② significance /sɪɡˈnɪfɪkəns/ *n.* 意义;重要性

③ efficient /ɪˈfɪʃnt/ *adj.* 有效率的;生效的

④ comprehensive /ˌkɒmprɪˈhensɪv/ *adj.* 综合的;广泛的

⑤ ingenuity /ˌɪndʒəˈnjuːəti/ *n.* 心灵手巧;聪明才智

⑥ incredible /ɪnˈkredəbl/ *adj.* 难以置信的;极好的

⑦ fortitude /ˈfɔːtɪtjuːd/ *n.* 刚毅;不屈不挠

⑧ foster /ˈfɒstə(r)/ *vt.* 促进;抚养

⑨ spectacular /spekˈtækjələ(r)/ *adj.* 壮观的;引人注目的

科技成就

中国高铁

China High-speed Railway

 导入语 Lead-in

　　自2008年8月1日中国第一条高铁线路——京津城际高铁开通运营以来，高速铁路在中国迅猛发展。截至2020年底，中国高铁总里程突破3.79万公里，覆盖中国80%的主要城市，是世界上最大的高速铁路网。高铁给人们的生活和工作带来了巨大变化，促进了沿路地区的经济发展，使相邻省份融为"一日生活圈"。中国高铁不仅为百姓出行提供了便利,更在世界舞台上大放异彩,向世界展示了一张靓丽的中国名片,散发着新时代中国制造的独特魅力。

文化剪影 Cultural Outline

China High-speed Railway is designed for speeds of 250～350 kilometers per hour. It is the world's longest and also the most **extensively**[①] used high-speed railway network.

中国高铁设计时速达250至350公里每小时,是世界上最长、也是使用最广泛的高速铁路网。

For its comfort, convenience, safety and **punctuality**[②], China High-speed Railway has not only changed the way people travel across the country, but also contributed to rapid economic growth.

中国高铁舒适、方便、安全、准时,不仅改变了人们的出行方式,也促进了经济的快速增长。

China High-speed Railway is becoming more and more important to national economy and social development. According to the railway development plan, China is building its "eight **horizontal**[③] and **vertical**[④]" high-speed railway network.

中国高铁对国民经济和社会发展越来越重要。根据铁路发展规划,中国正在建设"八横八纵"高速铁路网。

科技成就

佳句点睛 Punchlines

1. China High-speed Rains attract passengers from all income groups with fast speed and **favorable**⑤ fares.

中国高铁列车以其快捷的速度和优惠的票价吸引着各收入阶层的乘客。

2. The fastest Fuxing high-speed train can reach a speed of 350 kilometers per hour.

速度最快的复兴号高铁列车时速可达350公里每小时。

3. China's high-speed trains are expected to offer more smart services as the network continues to be upgraded with internet technology.

随着网络技术的不断升级，中国高铁列车有望提供更多的智能服务。

情景对话 Situational Dialogue

A: George, how long it takes from Beijing to Shanghai by train?

B: Emm... It's around 1,200 kilometers, similar to traveling from Seattle to San Francisco or from Paris to Florence.

A: Oh, it must be a long time.

B: Not really, it's only 4 hours and 18 minutes by the China's high-speed train running at 350 km/h.

A: It's so fast.

B: But China's train wasn't always this fast. When the People's Republic of China was founded in 1949, a train from Beijing to Shanghai would take nearly 37 hours. In 2008, China's first high-speed rail line from Beijing to Tianjin went into **operation**⑥. The line, which cuts travel time from two hours to 30 minutes, is a new beginning for China's railway development.

A: Is the train ticket expensive?

B: No, it keeps at a relatively low level, making it **accessible**⑦ to a large part of the population. Today, China **accounts for**⑧ more than two thirds of the world's total high-speed rail networks, covering 80 percent of major cities.

A: In this way, it is really convenient to go out and visit friends.

B: Yes, a broad range of travelers from different income levels would choose China high-speed railway for its comfort, convenience, safety and punctuality.

A: 乔治,坐火车从北京到上海要多长时间?

B: 嗯……大约1200公里的路程,相当于从西雅图到旧金山,或者从巴黎到佛罗伦萨。

A: 噢,那一定要坐很久吧。

B: 不一定,在时速350公里的中国高铁列车上,只需要4小时18分钟。

A: 这么快啊。

B: 但中国的高铁列车并不总是这么快。1949年中华人民共和国成立时,从北京到上海坐火车需要将近37个小时。2008年,中国第一条从北京到天津的高铁线投入运营,这条线路将行车时间从2小时缩短至30分钟,是中国铁路发展的新开端。

A: 车票贵吗?

B: 不贵,中国高铁的价格始终保持在一个较低的水平,使得大部分人都能享受福利。如今,中国高铁总里程占世界的2/3以上,覆盖了国内80%的主要城市。

A: 这样一来,外出访友真的很方便。

B: 是啊,因为它具有舒适、方便、安全和准时的优点,所以不同收入层次的旅客都会选择高铁出行。

生词注解 Notes

① extensively /ɪkˈstensɪvlɪ/ *adv.* 广阔地;广大地

② punctuality /ˌpʌŋktʃuˈælətɪ/ *n.* 准时;正确

③ horizontal /ˌhɒrɪˈzɒntl/ *adj.* 水平的;地平线的

④ vertical /ˈvɜːtɪkl/ *adj.* 垂直的;直立的

⑤ favorable /ˈfeɪvərəbl/ *adj.* 优惠的;有利的

⑥ operation /ˌɒpəˈreɪʃn/ *n.* 运行;操作

⑦ accessible /əkˈsesəbl/ *adj.* 易接受的;可理解的

⑧ account for 占……比例

南极科考

Antarctic[①] Expedition

导入语 Lead-in

南极是地球上最后一块尚未开发、尚未污染的洁净大陆，蕴藏着无数的科学之谜。自1984年中国开展南极科考以来，中国的南极科  考工作者历经艰难险阻，获得了大批极其宝贵的数据、资料和样品，取得了许多国际领先的科学成果，极大丰富了人类对南极的认识。目前，中国已经完成了三十六次南极科考之旅，设立了四个科考站，组建了一批重点实验室，培养了一支门类齐全、体系完备的科研队伍。在建的第五个科考站罗斯海新站预计于2022年竣工。作为一项造福人类的崇高事业，南极科考意义重大而深远。

 文化剪影　Cultural Outline

Since 1984, when the first scientific research team was sent to the Antarctic, China has **successively**② established four scientific research bases—the Great Wall Station, Zhongshan Station, Kunlun Station and Taishan Station, and achieved fruitful scientific results.

自1984年派出第一支前往南极地区的科考以来，中国先后成功建立了长城站、中山站、昆仑站和泰山站四个科研基地，并取得了丰硕的科学成果。

China's first domestically-made polar icebreaker "Xuelong-2" set sail on the 36th Antarctic expedition in October 2019 with the objectives to improve China's ability to cope with climate change and increase its participation in global **governance**③ in Antarctica.

中国首艘国产极地破冰船"雪龙2号"于2019年10月启程开展第三十六次南极科考，旨在提高中国应对气候变化的能力，增加对南极全球治理的进一步参与。

Developing a peaceful, stable and green Antarctic is in the common interest of all human beings, and is also our **commitment**④ to future generations. China will join the rest of the world in understanding, protecting and **utilizing**⑤ Antarctica.

发展一个和平、稳定、绿色的南极，符合全人类的共同利益，也是我们对子孙后代的承诺。中国将与世界各国一道共同认识、保护和利用南极。

 佳句点睛 Punchlines

1. Beautiful glaciers, lovely penguins and Antarctica's endless ice and snow has been attracting the eyes of the world.

美丽的冰川、可爱的企鹅和南极无边无际的冰雪世界一直吸引着全世界的目光。

2. China has started to construct its fifth Antarctic scientific expedition station, which is scheduled to be completed in four years to meet the needs of scientific expedition throughout the year.

中国已经开始建设第五个南极科考站，预计将在四年内建成，以满足全年的科学考察需求。

3. China has built its first domestically-made polar icebreaker "Xuelong-2", which will provide a **solid**⑥ guarantee for China's polar expeditions.

中国已经建成首艘国产极地破冰船"雪龙二号"，这将为极地探险提供坚实保障。

情景对话 Situational Dialogue

A: It's our great honor to invite you to share your Antarctica trip.

B: Thank you. Although it's already been two months since I returned, my excitement of stepping on Antarctica for the first time remains fresh in my memory.

A: Great! What preparation have you made before setting off for the region?

B: All members must study the rules and regulations of *The Antarctic Treaty System*.

A: *The Antarctic Treaty System?*

B: Yes, it's an international **pact**① **observed**② by fifty-three countries.

A: It's said that this expedition marks the end of the site selection process for China's fifth Antarctic scientific expedition station.

B: Yes, a suitable location was finally found on the west coast of the Ross Sea. Despite the harsh environment, it has significant research value due to its **unique**③ natural conditions.

A: And what's your main duty?

B: I was conducting geological investigations in the Ross Sea, including taking samples of the seabed.

A: Is there anything impressive?

B: Of course. We could sometimes see seals on the floating ice and whales coming out of the ocean while on the ship. I miss my life there and will keep track of information about the South Pole.

A: 我们很荣幸邀请您来分享您的南极之旅。

B: 谢谢。虽然我已经回来两个月了，但我对首次踏上南极时的兴奋之情依然记忆犹新。

A: 太棒了！你们出发之前做了哪些准备呢？

B: 所有成员都必须学习《南极条约》的规章制度。

A: 《南极条约》？

B: 是的，这是一项由五十三个国家共同遵守的国际公约。

A: 据说这次科考标志着中国第五个南极科考站选址工作的结束。

B: 是的，我们最终在罗斯海的西海岸找到了一个合适的地点。尽管那里环境恶劣，但独特的自然条件使其具有重要的研究价值。

A: 那您的职责是什么呢？

B: 我在罗斯海进行地质调查，包括采集海床样本。

A: 有什么印象深刻的事情吗？

B: 当然有。我们有时在船上可以看到浮冰上的海豹，还有从海里游出来的鲸鱼。我怀念那里的生活，并将持续关注有关南极的信息。

生词注解 Notes

① Antarctic /ænˈtɑːktɪk/ n. 南极洲；南极地区

② successively /sək'sesɪvlɪ/ adj. 相继地；接连地

③ governance /'gʌvənəns/ n. 管理；统治

④ commitment /kə'mɪtmənt/ n. 承诺；保证

⑤ utilize /'juːtəlaɪz/ vt. 利用；运用

⑥ solid /'sɒlɪd/ adj. (证据或信息)真实可靠的

⑦ pact /pækt/ n. 公约；协定

⑧ observe /əb'zɜːv/ vt. 遵守；观察

⑨ unique /juˈniːk/ adj. 独特的；唯一的

"蛟龙号"潜水器

"Jiaolong" Submersible[①]

导入语 Lead-in

"蛟龙号"潜水器是一艘由中国自行设计、自主集成研制的载人潜水器。"蛟龙号"长8.2米、宽3.0米、高3.4米,重约22吨,有效负载220公斤,设计潜水深度为7千米级。2012年7月,"蛟龙号"在马里亚纳海沟创造了下潜7062米的中国载人深潜纪录,也是世界同类作业型潜水器的最深下潜深度纪录,这意味着中国具备了载人到达全球99.8%以上海洋深处进行作业的能力。近年来,"蛟龙号"获得了大量海底样品和资料,在海洋地质、海洋生物等领域取得了丰硕的科学考察成果,对中国开发、利用深海资源具有十分重要的意义。

文化剪影 Cultural Outline

"Jiaolong" Submersible is a comprehensive scientific expedition vessel and China's first self-developed and specially designed manned submersible. In June 2012, it completed the deepest dive of 7,062 meters in the Mariana **Trench**② setting a world record.

"蛟龙号"潜水器是一艘综合科考船,也是中国第一艘自行研制、专门设计的载人潜水器。2012年6月,它在马里亚纳海沟完成了7062米的最深潜水,创下了世界纪录。

"Jiaolong" Submersible has successfully conducted 158 dives with a total voyage distance of 86,000 **nautical**③ miles, which demonstrates China's advancement in the development of deep-sea research.

"蛟龙号"潜水器成功下潜158次,总航程86000海里,标志着中国在深海研究发展方面取得的进步。

"Jiaolong" Submersible has two robotic arms that can take samples of seawater, **sediment**④, deep sea creatures and rocks as needed.

"蛟龙号"潜水器有两个机械臂,可以按照需求采集海水、沉积物、深海生物和岩石样本。

佳句点睛 Punchlines

1. "Jiaolong" Submersible is a Chinese manned deep-sea research submersible that can dive to a depth of over 7,000 meters.

"蛟龙号"潜水器是一艘中国载人深海研究潜水器,它可以潜到7000多米的深度。

2. Named after a **mythical**⑤ dragon, "Jiaolong" Submersible reached its deepest point of 7,062 meters in the Mariana Trench in June 2012.

2012年6月,以神话中的蛟龙命名的"蛟龙号"潜水器在马里亚纳海沟达到了7062米的最深下潜深度。

3. The deep-sea expedition of "Jiaolong" Submersible is tasked with exploring the adaptive and evolutionary mechanism of creatures in the depths of oceans.

"蛟龙号"潜水器的深海探险任务是探索深海生物的适应和进化机制。

情景对话 Situational Dialogue

A: It is reported that Chinese "Jiaolong" Submersible will have new mothership.

科技成就

B: Is its current mothership the "Xiangyanghong-9"?

A: Yes, it has carried "Jiaolong" for hundreds of dives since 2009 and is set to retire.

B: Has "Jiaolong" started its test dives at the National Deep-sea Center?

A: Yes. Once it passes all tests, "Jiaolong" will then **embark on**① its new ocean adventure with the carrier "Deepsea-1".

B: So, is "Deepsea-1" its new mothership?

A: Yes. It's designed especially for "Jiaolong", which boasts larger lab space, lower underwater noise levels and new **environmentally-friendly**② designs.

B: What has "Jiaolong" obtained during its diving?

A: Samples of rock, **sediment**③, deep-sea life and sea water were collected. And what's more, it was able to capture images of the ocean floor, which will offer **invaluable**④ information for scientists.

B: Really Great! China has implemented a strategy of becoming a **maritime**⑤ power by optimally utilizing its oceanic resources.

A: 据报道，中国的"蛟龙号"潜水器将拥有新的母船了。

B: 它现在的母船是"向阳红九号"吗？

A: 是的，自2009年以来，"向阳红九号"已经搭载"蛟龙号"潜水器数百次了，即将退役。

B: "蛟龙号"潜水器已经在国家深海中心开始试潜了吗？

A: 是的。一旦通过所有测试，它将与"深海一号"一起开始新的海底探险。

B：所以，"深海一号"就是它的新母船吗？

A：是的。"深海一号"是专门为"蛟龙号"潜水器设计的，它拥有更大的实验室空间、更小的水下噪音和全新的环保设计。

B："蛟龙号"潜水器在潜水过程中有哪些收获？

A：它采集了岩石、沉积物、深海生物和海水样本。另外，它还能够捕捉到海底的图像，为科学家提供宝贵的信息。

B：真了不起！通过优化利用海洋资源，中国实施了海洋强国战略。

生词注解 Notes

① submersible /səbˈmɜːsəbl/　*n.* 潜水器

② trench /trentʃ/　*n.* 地沟；沟渠

③ nautical /ˈnɔːtɪkl/　*adj.* 航海的；船员的

④ sediment /ˈsedɪmənt/　*n.* 沉积；沉淀物

⑤ mythical /ˈmɪθɪkl/　*adj.* 神话的；虚构的

⑥ embark on　从事；着手

⑦ environmentally-friendly /ɪnˌvaɪrənˈmentəli ˈfrendli/　*adj.* 环保的

⑧ sediment /ˈsedɪmənt/　*n.* 沉积；沉淀物

⑨ invaluable /ɪnˈvæljuəbl/　*adj.* 无价的；非常重要的

⑩ maritime /ˈmærɪtaɪm/　*adj.* 海运的；沿海的

"中国天眼"

"China's Eye of Heaven"

 导入语 Lead-in

　　500米口径球面射电望远镜被誉为"中国天眼",是我国具有自主知识产权、世界最大单口径、最灵敏的射电望远镜。"天眼"位于贵州省平塘县喀斯特地貌天坑中,由中国天文学家南仁东于1994年提出构想,历时22年落成启用。它能够接收137亿光年以外的电磁信号,并在未来10年至20年保持国际一流设备的地位,现已实现跟踪、漂移扫描、运动中扫描等多种观测模式,已经发现132颗优质的脉冲星候选体,其中93颗被确认为新发现的脉冲星。作为国家重大科技基础设施,"天眼"工程综合体现了中国高技术创新能力,并将在日地环境研究、国防建设和国家安全等方面发挥不可替代的作用。

 文化剪影 **Cultural Outline**

　　The 500 Aperture Spherical Telescope, named "The Eye of Heaven", is the world's largest and most **sensitive**① single-dish radio telescope, independently developed by Chinese scientists.

　　500米口径球面射电望远镜,又名"天眼",是由中国科学家自主研发的,世界上最大单口径、最灵敏的射电望远镜。

　　"China's Eye of Heaven" is located in a karst landform pit in Guizhou Province, southwestern China. Since the first discovery of pulsars in October 2017, 132 high-quality pulsar **candidates**② have been discovered, of which 93 have been confirmed.

　　"中国天眼"位于中国西南部贵州省的一个喀斯特地貌天坑中,自2017年10月首次发现脉冲星以来,已经发现了132颗优质的脉冲星候选体,其中93颗已经得到确认。

　　The realization of "China's Eye of Heaven" Project will bring golden opportunities for the development of related high-tech and promote economic and social development in the western region of China.

"中国天眼"项目的实施将为相关高新技术的发展带来良好的机遇,并促进西部地区的经济和社会发展。

佳句点睛 Punchlines

1. "China's Eye of Heaven" is very likely to make a breakthrough in the study of gravitational waves and general relativity.

"中国天眼"极有可能在引力波和广义相对论的研究上取得突破。

2. "China's Eye of Heaven" will **tremendously**① enhance the research abilities in the field of radio astronomy and other basic research area of China.

"中国天眼"的快速发展将极大提高中国射电天文等基础领域的研究能力。

3. The pulsar observation is an important task for "China's Eye of Heaven", which can be used to confirm the existence of gravitational radiation and black holes and help to solve many other major questions in physics.

脉冲星观测是"中国天眼"的一项重要任务,可以用来确认引力辐射和黑洞的存在，并帮助解决物理学中的许多重大问题。

 情景对话　Situational Dialogue

A: Do you know the "**Asteroid**① Nan Rendong"?

B: Yes. It's named after the Chinese scientist who has founded the Five-hundred-meter Aperture Spherical Telescope with approval from the International Astronomical Union.

A: Absolutely. Nan worked as the chief scientist in charge of its site selection and construction in 1994. He aimed to create the largest super-sensitive "ear" on earth in an attempt to seek distant sounds in the universe and **decode**② cosmic messages.

B: He has made a significant contribution to "China's Eye of Heaven" Project and **devoted to**③ his work without considering personal gains.

A: As a multi-science platform, the telescope will provide treasures to astronomers and bring prosperity to other research, such as space weather study, deep space exploration and national security.

B: It will also have an impact on many areas of astronomy and astrophysics around the world.

A: Although we can't predict what it will discover, the telescope may **profoundly**④ change our understanding of the universe.

B: It's expected that interesting and **exotic**⑤ stars will be discovered by FAST.

A: 你知道"南仁东星"吗?

B: 知道啊,它是经国际天文联合会批准,以中国科学家南仁东先生命名的,他是中国500米口径球面射电望远镜的建设者。

A: 是的。南仁东先生自1994年担任该项目的首席科学家以来,主要负责选址和施工。他的目标是创造地球上最大的超灵敏"耳朵",以捕捉宇宙中遥远的声音并解码宇宙信息。

B: 他在工作中从不考虑个人利益,为"中国天眼"做出了重要贡献。

A: 作为一个多学科平台,"中国天眼"将为天文学家提供宝贵的财富,并且促进其他研究领域如太空气候、深空探索和国家安全的繁荣发展。

B: "中国天眼"也会对全世界的天文和天体物理学领域产生影响。

A: 尽管我们无法预测它会发现什么,但这台望远镜可能会深刻改变我们对宇宙的认识。

B: 期待"中国天眼"发现奇趣的星体。

 生词注解 Notes

① sensitive /ˈsensətɪv/　*adj.* 敏感的;感觉的

② candidate /ˈkændɪdət/　*n.* 候选人;应试者

③ tremendously /trəˈmendəslɪ/　*adv.* 非常地;惊人地

④ asteroid /ˈæstərɔɪd/ n. (火星和木星间运行的)小行星

⑤ decode /diːˈkəʊd/ vt. 解码

⑥ devote to 将……奉献给；把……专用于

⑦ profoundly /prəˈfaʊndlɪ/ adv. 深刻地；极度地

⑧ exotic /ɪgˈzɒtɪk/ adj. 奇特的；异域的

科技成就

"天河二号"超级计算机

"Tianhe-2" Supercomputer

 导入语　Lead-in

"天河二号"超级计算机是由国防科技大学研制的,以峰值计算速度每秒5.49亿亿次、持续计算速度每秒3.39亿亿次双精度浮点运算的优异性能,成为2013年全球运算速度最快的超级计算机。"天河二号"由170个机柜组成,占地面积720平方米,其存储容量相当于存储每册10万字的图书600亿册。"天河二号"运算1小时,相当于13亿人同时用计算器计算1000年。随着"天河二号"应用领域的不断扩展,中国的超级计算机已经日益显示出作为国家重要基础设施的强大支撑作用。截至2019年11月18日,"天河二号"在全球超级计算机500强榜单中位列第四。

文化剪影 Cultural Outline

"Tianhe-2" Supercomputer was designed by National University of Defense Technology and **crowned**① as the fastest supercomputer in the world in June 2013.

"天河二号"由国防科技大学设计,2013年6月被评为"世界上最快的超级计算机"。

"Tianhe-2" Supercomputer has been widely used in the fields of **simulation**②, analysis and national security.

"天河二号"超级计算机已经广泛应用于模拟、分析和国防安全等领域。

"Tianhe-2" set new world records for both peak speed and sustained speed, focusing on scientific engineering computing and cloud computing, **innovatively**③ developing **heterogeneous**④ fusion architecture and improving software **compatibility**⑤ and ease of programming.

"天河二号"的峰值速度和持续速度都创造了新的世界纪录,其主打科学工程计算,兼顾云计算,创新发展了异构融合体系结构,提高了软件兼容性和易编程性。

佳句点睛 Punchlines

1. "Tianhe-2" won the championship for the third time in a row nearly twice as fast as the American "Titan".

"天河二号"以比美国"泰坦"快近一倍的速度连续第3次获得冠军。

2. "Tianhe-2" Supercomputer was honored as the world fastest supercomputer according to the Top 500 international high-performance computer for 6 **consecutive**① times.

天河二号超级计算机连续6次被国际高性能计算机500强榜单评为"世界上最快的超级计算机"。

3. The cosmic Neutrino numerical simulation team has successfully performed a 3 trillion particle number Neutrino and dark matter **cosmological**② numerical simulation on the "Tianhe-2", revealing the long evolution of the universe after the Big Bang.

宇宙中微子数值模拟团队在"天河二号"上成功进行了30000亿粒子数中微子和暗物质的宇宙学数值模拟,揭示了宇宙大爆炸后的漫长演化进程。

情景对话 Situational Dialogue

A: I went to the National Supercomputing Center in Guangzhou last weekends.

B: Is the "Tianhe-2" there? I know it defeated America's "Titan".

A: Yes, "Tianhe-2" was the world's fastest supercomputer according to the Top 500 lists for June 2013. It was designed and developed by a team of 1,300 scientists and engineers.

B: It's said that the Top 500 ranks and details the 500 most powerful supercomputers in the world.

A: Yes. The project was started in 1993 and publishes an updated list twice a year. It is important in that it **tracks**® historical performance of supercomputers and predicts their future development.

B: And why is "Tianhe-2" located in the south?

A: Because the warmer weather with higher temperature in the south could reduce the electricity **consumption**® by about 10% compared to the north.

B: It's still leading the world in terms of calculation capacity. I'd like to see it.

A: 上周末我去了位于广州的国家超级计算机中心。

B: "天河二号"就在那里,对吗? 我知道它打败了美国的"泰坦"。

科技成就

A: 是啊。根据2013年6月的世界超级计算机500强名单,"天河二号"是当时最快的超级计算机,是由1300名科学家和工程师组成的团队设计开发的。

B: 据说这个500强项目对世界上500台最强大的超级计算机进行了排序和详细记录。

A: 是的。该项目始于1993年,每年两次发布最新的名单。这对于追踪超级计算机的历史性能并预测它们的未来发展很重要。

B: "天河二号"为什么在南方呢?

A: 因为相比于北方,南方较温暖的天气和较高的温度可以减少约10%的电力消耗。

B: 它在计算方面仍然位于世界前列。我想去看看。

生词注解 Notes

① crown /kraʊn/ *vt.* 加冕;表彰

② simulation /ˌsɪmjuˈleɪʃn/ *n.* 模拟;仿真

③ innovatively /ˈɪnəveɪtɪvli/ *adv.* 独创性地;创新地

④ heterogeneous /ˌhetərəˈdʒiːniəs/ *adj.* (数学)不纯一的;参差的

⑤ compatibility /kəmˌpætəˈbɪləti/ *n.* 兼容性;相容性

⑥ consecutive /kənˈsekjətɪv/ *adj.* 连贯的;连续不断的

⑦ cosmological /ˌkɒzməˈlɒdʒɪkl/ *adj.* 宇宙论的;宇宙哲学的

⑧ track /træk/ *vt.* 追踪;循路而行

⑨ consumption /kənˈsʌmpʃn/ *n.* 消费;消耗

"神威·太湖之光"超级计算机

"Sunway TaihuLight" Supercomputer

导入语 Lead-in

"神威·太湖之光"超级计算机是由国家并行计算机工程技术研究中心研制,世界首台峰值计算速度超过10亿亿次的超级计算机。这台计算机全部采用国产芯片——"神威

26010"众核处理器,由40个运算机柜和8个网络机柜组成,占地面积605平方米。其峰值运算速度达到每秒12.54亿亿次,持续运算速度每秒9.3亿亿次,一分钟计算能力相当于全球72亿人同时用计算器不间断计算32年。在2016年世界超算大会上,"神威·太湖之光"超级计算机登顶榜单之首。它对我国科技创新和经济发展具有十分重要的作用,其应用领域涉及天气气候、航空航天、先进制造和新材料等方面。

文化剪影 Cultural Outline

"Sunway TaihuLight" Supercomputer was developed by China's National Research Center of **Parallel**① Computer Engineering and Technology, and installed at the National Supercomputing Center in Wuxi City, Jiangsu Province.

"神威·太湖之光"超级计算机由中国国家并行计算机工程技术研究中心开发,安装在江苏省无锡市的国家超级计算中心。

"Sunway TaihuLight" Supercomputer was again at the top of a **biannual**② ranking of the world's 500 fastest supercomputers in November 2015, maintaining the lead as the No. 1 system for four consecutive years.

在2015年11月的世界500台最快超级计算机半年度排名中,"神威·太湖之光"超级计算机再次位居榜首,连续四年蝉联世界第一。

One of Sunway's original characteristics is domestic technology. Sunway TaihuLight Supercomputer was built by entirely using processors designed and made in China.

国产化是神威公司的一大特色,"神威·太湖之光"超级计算机完全使用中国设计和制造的处理器。

佳句点睛　Punchlines

1. "Sunway TaihuLight" Supercomputer, installed at the National Supercomputing Center in Wuxi, holds the third position with 93.0 **petaflops**③.

安装在无锡国家超级计算中心的"神威·太湖之光"超级计算机，以每秒93千万亿次的速度位居世界第三。

2. "Sunway TaihuLight" Supercomputer, built with Chinese technology, has contributed significantly to research benefiting the Chinese people.

中国制造的"神威·太湖之光"超级计算机为惠民科研做出了重大贡献。

3. With "Sunway TaihuLight" Supercomputer as the main computing power, a Chinese team designed a software that calculates the flow of Earth's atmosphere, winning the 2016 Gordon Bell Prize, honored as the Nobel Prize in Global Computer Science.

用"神威·太湖之光"超级计算机进行计算的中国团队设计了一款计算地球大气流量的软件，并赢得了2016年度戈登·贝尔奖，该奖项被誉为全球计算机科学界的诺贝尔奖。

科技成就

情景对话 Situational Dialogue

A: Professor Hu, why do supercomputers matter?

B: Due to their unparalleled data-processing capability, the supercomputer can be widely used in fields like Artificial Intelligence, medical research and national security.

A: So how fast is a supercomputer?

B: Take China's "Sunway TaihuLight" Supercomputer as an example, its peak performance can reach 125,000 teraflops, meaning it's 125,000 times more powerful than the best personal computer **available**④ at stores.

A: It's a **dazzling**⑤ speed.

B: What's more, it can benefit humanity, even life-saving ones. For example, it is able to detect and warn of an oncoming Tsunami 10 to 20 minutes ahead of time, or produce a one-hundred-year Earth's climate simulation in 30 days.

A: Incredible! China has made such huge progress in technological **innovation**⑥.

B: However, China's computing industry does not rest on its **laurel**⑦. The next-generation "Sunway E-class" is under construction, which is expected to be eight times faster than "Sunway TaihuLight".

A: 胡教授,为什么超级计算机那么重要?

B: 由于其无与伦比的数据处理能力, 超级计算机可以广泛应用于人工智能、医学研究和国家安全等领域。

A: 那超级计算机有多快呢?

B: 以中国的"神威·太湖之光"超级计算机为例, 它的最高运算速度可以达到125000万亿次浮点运算, 这意味着它的运算能力是市面上最好的个人电脑的125000倍。

A: 这个速度太惊人了。

B: 此外, 它可以造福人类甚至拯救生命。比如, 它能够提前10到20分钟探测并预警即将到来的海啸, 或者在30天内完成100年间的地球气候模拟。

A: 太不可思议了, 中国在技术创新方面取得了如此巨大的进步。

B: 然而, 中国的计算机产业并不满足于现状。下一代"神威E级"正在建设中, 预计其速度将比"神威·太湖之光"快8倍。

生词注解　Notes

① parallel /ˈpærəlel/　*adj.* 平行的; 类似的

② biannual /baɪˈænjuəl/　*adj.* 一年两次的

③ petaflop /ˈpetəˌflɒp/　*n.* 千万亿次

④ available /əˈveɪləbl/　*adj.* 可获得的; 有空的

⑤ dazzling /ˈdæzlɪŋ/　*adj.* 眼花缭乱的; 耀眼的

⑥ innovation /ˌɪnəˈveɪʃn/　*n.* 创新; 革新

⑦ laurel /ˈlɒrəl/　*n.* 桂冠; 殊荣

北斗卫星导航系统

Beidou Navigation Satellite System

 导入语 Lead-in

北斗卫星导航系统是中国着眼于国家安全和经济社会发展需要，自行研制、独立运行的全球卫星导航系统，为用户提供全天候、全天时、高精度的定位、导航和授时服务。该系统由空间段、地面段和用户段三部分组成，是国家重要的空间基础设施。北斗卫星导航系统已经逐步融入现代化建设和百姓的日常生活。未来，北斗卫星导航系统将继续提升服务性能，扩展服务领域，增强运行能力，持续、稳定地推动全球卫星导航事业发展，更好地服务全球、造福人类。

文化剪影 Cultural Outline

Beidou Navigation Satellite System is a global satellite navigation system built and operated independently, providing users with high-precision and reliable positioning, navigation and timing services around the clock.

北斗卫星导航系统是中国自主研制、独立运行的全球卫星导航系统,为用户提供全天候、高精度、可靠的定位、导航和授时服务。

Named after the Chinese term for the Big Dipper **constellation**[①] and constructed in the 1990s, Beidou Navigation Satellite System is the fourth satellites navigation system in the world to cover the whole planet, after the United States' GPS, Russia's GLONASS and European Union's Galileo.

以北斗七星命名、建于20世纪90年代的北斗卫星导航系统,仅次于美国的全球定位系统、俄罗斯的格洛纳斯和欧盟的伽利略,是世界上第四个覆盖全球的导航卫星系统。

Beidou Navigation Satellite System has produced significant economic and social outcomes, fields including smart city, meteorological environment and **precise**[②] machine control are all its **beneficiaries**[③].

北斗卫星导航系统带来了显著的经济效益和社会效益,在智慧城市、气象环境、精密机械控制等领域都创造了颇多收益。

佳句点睛　Punchlines

1. Beidou Navigation Satellite System has become one of the most important achievements from China's reform and opening-up in the past forty years, and has been widely used in various industries.

北斗卫星导航系统已成为中国改革开放四十年来最重要的成果之一,广泛应用于各个行业。

2. Beidou Navigation Satellite System is capable of providing positioning services with an **accuracy**④ within 10 meters, while accuracy in the Asia-Pacific region can be less than five meters.

北斗卫星导航系统可提供精度在10米以内的定位服务,在亚太地区的精度甚至可以小于5米。

3. By 2035, a more **ubiquitous**⑤, integrated and smarter positioning, navigation and timing system with Beidou Navigation Satellite System as the core is expected to be established.

到2035年,将形成以北斗卫星导航系统为核心,更加普适、集成、智能化的定位、导航和授时系统。

情景对话 Situational Dialogue

A: Taxi, we'd like to go to Tiantan Park.

B: Get in, please.

A: Wow, your navigation system is so convenient.

B: Yes, it's Beidou Navigation Satellite System. About half of the taxis in Beijing are now **equipped with**⑥ it.

A: Is it one of the four major global positioning and navigation systems?

B: Yes. It maintains stable operation, **enhanced**⑦ performance and reliable services.

A: Why is it called Beidou?

B: It is named after the Big Dipper asterism, which was given by ancient Chinese **astronomers**⑧ to the seven brightest stars and historically used in navigation to locate the North Star. As such, the name "Beidou" also serves as a **metaphor**⑨ for the purpose of the satellite navigation system.

A: The application of Beidou System in transportations have made a **digitalized**⑩ and mobile China even more possible.

B: Yes. It has been used in the construction of "One Belt and One Road", providing global users with more efficient and reliable navigation services.

科技成就

A：出租车，我们去天坛公园。

B：请上车。

A：哇，您的导航系统真方便。

B：是的，这是北斗卫星导航系统。北京一半以上的出租车上都安装了呢。

A：它是全球四大定位导航系统之一吗？

B：是啊，它运行稳定、性能提升、服务可靠。

A：为什么叫北斗呢？

B：它是以北斗七星星群的名字命名的。北斗七星是中国古代天文学家给天上最亮的七颗星星取的名字，历史上曾用其来定位北极星。因此，"北斗"这个名字也隐含着将其用作卫星导航系统的目的。

A：北斗系统在交通运输领域的应用，让中国的社会发展更加数字化、移动化。

B：是啊，北斗系统已经应用于"一带一路"建设，为全球用户提供更加高效、可靠的导航服务。

生词注解 Notes

① constellation /ˌkɒnstəˈleɪʃn/　*n.* 星群；星座

② precise /prɪˈsaɪs/　*adj.* 精确的；明确的

③ beneficiary /ˌbenɪˈfɪʃəri/　*n.* 受益人；封臣

④ accuracy /ˈækjərəsi/　*n.* 精确度；准确性

⑤ ubiquitous /juːˈbɪkwɪtəs/　*adj.* 普遍存在的；无所不在的

⑥ be equipped with 以……装备；配备
⑦ enhanced /ɪnˈhɑːnst/ adj. 提高的；增强的
⑧ astronomer /əˈstrɒnəmə(r)/ n. 天文学家
⑨ metaphor /ˈmetəfə(r)/ n. 暗喻；隐喻
⑩ digitalized /ˈdɪdʒɪtəlaɪzd/ adj. 数字化的

科技成就

"墨子号"量子通信卫星

"Mozi" Quantum[①] Communication Satellite

导入语 Lead-in

"墨子号"量子通信卫星是由中国科学院国家空间科学中心研制的世界首颗量子科学实验卫星。2016年8月16日1时40分,"量子号"卫星于酒泉卫星发射中心发射升空。为了纪念中国古代科学家墨子在物理学尤其是光学领域的突出成就,该卫星命名为"墨子号"。"量子号"的成功发射和在轨运行,有助于中国在量子通信技术实用化整体水平上保持和扩大国际领先地位,实现国家信息安全和信息技术水平的跨越式提升,推动中国科学家在量子科学前沿领域取得重大突破,对于中国空间科学卫星的可持续发展具有重大意义。

文化剪影　Cultural Outline

"Mozi" Quantum Communication Satellite is the first satellite **deployed**② successfully in near-Earth space for quantum scientific experiments in the world, in which experimental study on the **fundamental**③ questions of quantum **mechanics**④ can be done.

"墨子号"量子通信卫星是世界上第一颗成功部署在近地空间进行量子科学实验的卫星,可以对量子力学的基本问题进行实验研究。

The quantum communications satellite, named after the ancient Chinese philosopher and scientist Mozi, who was the first person in the world to conduct **optical**⑤ experiments.

量子通信卫星以中国古代哲学家和科学家墨子的名字命名,因为墨子是世界上第一位进行光学实验的人。

The "Mozi" Quantum Communications Satellite team laid the groundwork for the future of an absolutely secure communications network, winning the 2018 top science prize—the Newcomb Cleveland Prize.

"墨子号"量子通信卫星团队为建立未来绝对安全的通信网络奠定了基础,因此赢得了2018年最高科学奖——克利夫兰奖。

科技成就

佳句点睛　Punchlines

1. "Mozi" Quantum Communication Satellite was launched from the Jiuquan Satellite Launch Center at 1:40 Beijing time on August 8, 2016, marking a new era of space science for China.

"墨子号"量子通信卫星于2016年8月16日1时40分从酒泉卫星发射中心发射,标志着中国太空科学迈入了新时代。

2. The scientific mission of Quantum Communication Satellite Project is to build a satellite which is equipped with a set of payloads for generating, **transmitting**① and receiving quantum signals.

量子通信卫星项目的科学使命是打造一颗装有一套有效载荷以产生、发射和接收量子信号的卫星。

3. Quantum Satellite is designed to establish "hack-proof" communications between parties by transmitting uncrackable keys from space to ground stations.

量子卫星通过将无法破解的密钥从太空传输到地面站建立"防黑客"通信。

情景对话　Situational Dialogue

A: Do you know the Newcomb Cleveland Prize?

B: Yes, it was established by the American Association for the Advancement of Science. The most academically valuable and **influential**① research papers published in *Science* have been awarded each year since 1923.

A: The 2018 Newcomb Cleveland Prize went to a team of 34 Chinese physicists **dedicated to**⑧ quantum communication. This is the first time that a Chinese team has won the prize with homegrown research.

B: What have they accomplished?

A: They have launched the world's first quantum satellite named "Mozi" in August 2016. As a result, it laid the groundwork for quantum communication of the future.

B: What is quantum communication?

A: It is an absolutely safe way of communication, which can neither be separated nor **duplicated**⑨. So it is impossible to wiretap, **intercept**⑩ or crack the information it transmits.

B: It sounds so cool.

A: 你知道克利夫兰奖吗？

B: 知道啊，它是由美国科学促进会设立的奖项。从1923年开始，每年颁发给在《科学》杂志上发表的最有学术价值和影响力的论文。

A: 2018年的克利夫兰奖授予了一个由34名中国物理学家组成的团队，他们致力于量子通信研究。这是中国团队首次凭借自主研究成果获得该奖项。

B: 他们取得了怎样的成就?

A: 他们于2016年8月发射了世界上第一颗量子卫星"墨子号"。这为未来的量子通信奠定了基础。

B: 什么是量子通信?

A: 它是一种绝对安全的通信方式,因为量子既不能分离也不能复制,所以不可能被窃听、拦截或破解所传输的信息。

B: 听起来很酷啊。

生词注解 Notes

① quantum /ˈkwɒntəm/ n. 量子

② deploy /dɪˈplɔɪ/ vt. 部署;配置

③ fundamental /ˌfʌndəˈmentl/ adj. 基本的;根本的

④ mechanics /mɪˈkænɪks/ n. 力学;机械学

⑤ optical /ˈɒptɪkl/ adj. 光学的;视觉的

⑥ transmit /trænzˈmɪt/ vt. 传输;发射

⑦ influential /ˌɪnfluˈenʃl/ adj. 有影响的;有势力的

⑧ dedicate to 献身;把(时间、精力等)用于……

⑨ duplicate /ˈdjuːplɪkeɪt/ vt. 复制;重复

⑩ intercept /ˌɪntəˈsept/ vt. 拦截;截断

现代科技成就 第二部分

"悟空号"暗物质粒子探测卫星
"Wukong" Dark Matter Particle[①] Explorer

导入语 Lead-in

"悟空号"暗物质粒子探测卫星是中国第一个空间高能粒子探测器,也是目前世界上观测能段范围最宽、能量分辨率最高的暗物质探测卫星。2015年12月17日8时12分,中国在酒泉卫星发射中心成功发射了暗物质粒子探测卫星。经过全球征名活动,最终将其命名为"悟空",寓意火眼金睛的孙悟空寻找难以察觉的暗物质踪影。"悟空号"暗物质粒子探测卫星将在太空中探寻暗物质存在的证据,研究暗物质特性与空间分布规律,有望在物理学与天文学前沿带来新的重大突破。

文化剪影　Cultural Outline

"Wukong" Dark Matter Particle Explorer is China's first astronomical satellite dedicated to the indirect detection of dark matter particles and the study of high-energy astrophysics. It has the widest observation **spectrum**② and the highest energy resolution in the world at present.

"悟空号"暗物质粒子探测卫星是中国第一颗致力于间接探测暗物质粒子和研究高能天体物理学的天文卫星。它是目前世界上观测能段范围最宽、能量分辨率最优的暗物质粒子探测卫星。

"Wukong" Dark Matter Particle Explorer is designed to carry out high-energy **electron**③ and Gama ray detection in space in search of evidence of the existence of dark matter and for research on dark matter properties and **distribution**④ rules in space.

"悟空号"暗物质粒子探测卫星用于在太空中进行高能电子和伽马射线的探测,寻找暗物质存在的证据,研究暗物质的性质和空间分布规律。

Dark Matter Particle Explorer is nicknamed "Wukong" after the Monkey King, who is the hero in the *Journey to the West*, a classic Chinese novel. Literally, "wu" means comprehension or understanding and "kong" means **void**⑤, relating to the undiscovered nature of dark matter.

暗物质粒子探测卫星以美猴王的名字"悟空"命名,美猴王是中国古典小说《西游记》中的主人公。从字面上看,"悟"的意思是领悟或理解,"空"的意思是虚空,这都与暗物质未被发现的特质有关。

佳句点睛 Punchlines

1. The main scientific objective of "Wukong" Dark Matter Particle Explorer is to detect electrons and photons with unprecedented energy resolution in order to identify possible dark matter signatures.

"悟空号"暗物质粒子探测卫星的主要科学目标是用前所未有的能量分辨率探测电子和光子,以识别可能存在的暗物质信号。

2. "Wukong" Dark Matter Particle Exploring is a space telescope used for the detection of high energy gamma rays, electrons and **cosmic**⓵ ray ions, to aid in the search for dark matter.

"悟空号"暗物质粒子探测卫星是一种太空望远镜,用于探测高能伽马射线、电子和宇宙射线离子,以帮助寻找暗物质。

3. With its excellent photon detection capability, "Wukong" Dark Matter Particle Exploring is well placed for new discoveries in high-energy ray astronomy as well.

凭借其卓越的光子探测能力,"悟空号"暗物质粒子探测卫星为高能射线天文学的新发现奠定了良好基础。

科技成就

情景对话 Situational Dialogue

A: The universe is composed of 73% dark energy, 23% dark matter and 4% **baryonic**⑦ matter.

B: So the dark energy and dark matter have account for more than 90%. What are they?

A: They play a **crucial**⑧ role in the evolution of the universe, and may decide the fate of the universe. What's more, the dark energy and dark matter are referred to by scientists as the "two clouds" **obscuring**⑨ astronomy in the 21st century.

B: How could we detect them?

A: "Wukong" Dark Matter Particle Explorer is the first high-energy detector satellite of China, whose physics goal is to find evidence of the existence of dark matter particles.

B: It must has important implications in promoting innovative development of space science in China.

A: Yes. Nowadays, the research on the composition of universe from home and abroad has been **flourishing**⑩.

B: It is predicted that within the coming ten years it will be a golden age for studying and detecting dark energy and dark matter.

A: 宇宙是由73%的暗能量、23%的暗物质和4%的重子物质组成的。

B: 这么说,暗能量和暗物质占了90%以上。它们是什么呢?

A: 它们在宇宙演化过程中起着至关重要的作用,很可能决定着宇宙的命运。而且,暗能量和暗物质被科学家们称为21世纪遮蔽天文学的"两团乌云"。

B: 我们怎样才能观测到它们呢?

A: "悟空号"暗物质粒子探测卫星是我国第一颗高能探测卫星,其物理目标就是寻找暗物质粒子存在的证据。

B: 那它一定对促进中国空间科学的创新发展具有重要的意义。

A: 是的。目前,国内外对宇宙组成的研究方兴未艾。

B: 可以预见,未来十年将是研究和探测暗能量和暗物质的黄金时代。

生词注解　Notes

① particle /ˈpɑːtɪkl/　*n.* 颗粒;粒子

② spectrum /ˈspektrəm/　*n.* 光谱;范围

③ electron /ɪˈlektrɒn/　*n.* 电子

④ distribution /ˌdɪstrɪˈbjuːʃn/　*n.* 分布;供应

⑤ void /vɔɪd/　*adj.* 空的;无效的

⑥ cosmic /ˈkɒzmɪk/　*adj.* 宇宙的

⑦ baryonic /ˌbærɪˈɒnɪk/　*adj.* 重子的

⑧ crucial /ˈkruːʃl/　*adj.* 重要的;决定性的

⑨ obscure /əbˈskjʊə(r)/　*vt.* 掩盖;使……模糊不清

⑩ flourishing /ˈflʌrɪʃɪŋ/　*adj.* 繁荣的;盛行的

国产大飞机 C919

Domestically-made Airplane C919

C919是中国首款严格按照最新国际适航标准和国际民航规章自行研制，具有自主知识产权的干线民用飞机。C是中国英文名称"China"的首字母，第一个数字"9"寓意天长地久，"19"代表的是这款大型客机的最大载客量为190座。C919大飞机专为短程到中程的航线设计，确保安全性，突出经济性，提高可靠性，改善舒适性，强调环保性。2017年5月5日，C919大飞机完成首飞。它凝聚着中国数十万科研人员的心血，承载着几代中国人的航空梦想，体现了中国航空工业水平的巨大进步。

文化剪影 Cultural Outline

As China's first self-developed trunk jetliner, C919 large airplane is designed according to the newest Standard Airworthiness Certification and has conducted its successful **maiden**① flight at Shanghai Pudong International Airport on May 5, 2017.

作为中国自主研制的第一架干线喷气式客机，C919大飞机根据最新适航标准设计，于2017年5月5日在上海浦东国际机场成功进行了首次飞行。

The "C" in the name stands for "China" while "9" symbolizes "everlasting and **enduring**② " in Chinese culture, and "19" represents the maximum capacity 190 seats.

飞机名字中的"C"代表"中国"，而"9"在中国文化中象征着"永远"，"19"代表最大载客量为190座。

C919 has improved the **aviation**③ landscape in China and given a significant **boost**④ to China's aerospace industry.

C919改善了中国的航空环境，极大地推动了中国航空工业的发展。

佳句点睛　Punchlines

1. The Commercial Aircraft Corporation of China is the manufacturer of the C919, China's first domestically designed and built large passenger airplane.

中国商用飞机有限责任公司是C919大飞机的制造商，C919是中国第一架自主设计和制造的大型客机。

2. Shanghai-based China Eastern Airlines **confirmed**① that it will be the first customer to take delivery of the C919.

总部位于上海的中国东方航空公司证实，它将是第一个接收C919的客户。

3. With a standard range of 4,075 kilometers, the C919 is **comparable**② with the updated Airbus 320 and Boeing's new generation 737, signaling China's entry into the global aviation market.

C919大飞机的标准航程为4075公里，可与更新后的空客320和新一代波音737相媲美，标志着中国挺进全球航空市场。

 情景对话　Situational Dialogue

A: How does the Chinese-made passenger airplane C919 differ from its competitors?

B: The C919 is China's first domestically-built large passenger airplane and it was designed with the world's most **advanced**[7] technology. Main rivals on the market include the Airbus 320 and Boeing 737.

A: When has the C919 conducted its maiden flight?

B: In May 2017, the first C919 took to the skies, marking a milestone for the Commercial Aircraft Corporation of China, the Shanghai-based manufacturer of the C919.

A: It is said that the C919 will enter into a new phase of **intensive**[8] test flights.

B: Yes. These flight tests are a key verification phase for the aircraft model to **validate**[9] its design and performance, especially its safety aspects.

A: How competitive will the C919 be?

B: To date, it has received 815 orders from 28 domestic and foreign customers.

A: 中国制造的C919大飞机和它的竞争对手有什么不同？

B: 它是中国第一架国产大型客机，采用了世界上最先进的技术。市场上的竞争对手主要有空客320和波音737。

A: C919大飞机什么时候完成了首飞？

B: 2017年5月，第一架C919飞上天空，这对于它的制造商——位于上海的中国商用飞机有限责任公司来说是具有里程碑意义的。

A: 有消息称，C919将进入密集试飞的新阶段。

B: 是的。这些飞行测试是对飞机模型的设计和性能,特别是安全方面进行的重要验证。

A: C919大飞机的竞争力如何?

B: 到目前为止,飞机已经收到来自28个国内外客户的815份订单。

生词注解　Notes

① maiden /ˈmeɪdn/　*adj.* (航行、飞行)首次的;未婚的

② enduring /ɪnˈdjʊərɪŋ/　*adj.* 持久的;能忍受的

③ aviation /ˌeɪviˈeɪʃn/　*n.* 航空;飞机制造业

④ boost /buːst/　*vt.* 促进;增加

⑤ confirm /kənˈfɜːm/　*v.* 证实;确认

⑥ comparable /ˈkɒmpərəbl/　*adj.* 可比较的;同等的

⑦ advanced /ədˈvɑːnst/　*adj.* 先进的;高级的

⑧ intensive /ɪnˈtensɪv/　*adj.* 集中的;加强的

⑨ validate /ˈvælɪdeɪt/　*vt.* 验证;使……生效

5G 领跑世界

5G Leading the World

导入语 Lead-in

5G，全称"The 5th Generation Mobile Communication Technology"，即第5代移动通信技术。5G是最新一代蜂窝移动通信技术，具有全新的网络架构，提供峰值速率为10 Gbps以上的带宽、毫秒级的时延和超高密度的连接。互联网的蓬勃发展是5G移动通信的主要驱动力。2019年11月1日，我国正式上线5G商用套餐，成为当代世界信息化发展的领跑者。可以预见，5G将实现网络性能的整体跃升，并与物联网、大数据、云计算、人工智能、区块链等技术深度融合，实现

科技成就

万物互联互通信息传播发展提升的新飞跃。

文化剪影 Cultural Outline

5G is the latest generation of **cellular**[①] mobile communications, with data **transmission**[②] rates far higher than those of previous cellular networks, reaching up to 10 Gbps, 100 times faster than the previous 4G LTE cellular network, and faster response times to provide services not only for mobile phones, but also for home and office networks.

5G是最新一代的蜂窝移动通信,数据传输速率远远高于先前的蜂窝网络,最高可达10 Gbps,比先前的4G LTE蜂窝网络快100倍,具有更快的响应时间,不仅可以为手机提供服务,还可以为家庭和办公网络提供服务。

In the 5G age, almost everything can be **digitized**[③] and shared through various **terminals**[④] such as smart phones, cameras, cars, and **sensors**[⑤].

在5G时代,几乎所有东西都可以数字化,并通过智能手机、摄像头、汽车和传感器等各种终端进行共享。

A **fusion**[⑥] of cutting-edge network technologies and rigorous research, 5G is expected to pave the way towards a smarter and more connected world.

5G融合了尖端网络技术和严谨的研究,有望为一个更智能、更互联的世界铺平道路。

佳句点睛 Punchlines

1. The most remarkable features of the 5G network is its high speed and low **latency**①.

5G网络最显著的特点是高速度和低延迟。

2. 5G doesn't only offer faster speed, but also enable massive connections.

5G不仅能提供更快的速度,而且支持大规模连接。

3. Combined with **artificial**② intelligence, the 5G network will make online robots, virtual reality and other interactive services more widespread.

5G网络与人工智能相结合,将使在线机器人、虚拟现实和其他互动服务更加普及。

情景对话 Situational Dialogue

A: I attended the World 5G Convention in Beijing last week. Its theme is that "5G changes the world and creates the future".

B: Wow, is there anything interesting to share?

A: Of course. Sitting in the **simulated**③ cabin at the exhibition

187

hall, I could remotely control a self-driving car in Chongqing, about 1,500 kilometers away. 5G could also allow two musicians dozens of kilometers apart to play together in the same place.

B: So amazing!

A: 5G can achieve more than these functions. 5G is going to make our lives and work more intelligent and convenient in the near future.

B: Compared with previous generations of communication technologies, what are the characteristics of 5G?

A: It has three features as follows: ultra-high speed, lower latency and wider network **connectivity**[①]. So we call 5G a new beginning, a turning point and a revolution.

B: China is indeed a pioneer in the research and launch of a new generation of ultra-high-speed mobile networks.

A: 我参加了上周在北京举办的世界5G大会。它的主题是"5G改变世界,5G创造未来"。

B: 哇,有什么有趣的事情分享吗?

A: 当然有。我坐在北京展览馆的模拟机舱里,就可以遥控远在1500公里外的重庆市的无人驾驶汽车。5G也能使两个相距数十公里的音乐家在同一个地方表演合奏。

B: 太震撼了!

A: 5G能实现的功能远不止这些。在不久的将来,5G会让我们的生活更智能、更方便。

B: 相比于前几代通信技术,5G有什么特点呢?

A：它有三大特点：超高速、低延迟和更广泛的网络连接能力。因此，我们称5G是一个新的开始，是一个转折点，是一场革命。

B：中国不愧是研究和推出新一代超高速移动网络的先行者。

生词注解 Notes

① cellular /ˈseljələ(r)/　*adj.* 多孔的；细胞的

② transmission /trænzˈmɪʃn/　*n.* 传送；传播

③ digitized /ˈdɪdʒɪtaɪzd/　*adj.* 数字化的

④ terminal /ˈtɜːmɪnl/　*n.* 终端机；终点站

⑤ sensor /ˈsensə(r)/　*n.* 传感器

⑥ fusion /ˈfjuːʒn/　*n.* 融合；熔化

⑦ latency /ˈleɪtənsɪ/　*n.* 延迟；潜伏

⑧ artificial /ˌɑːtɪˈfɪʃl/　*adj.* 人工的；人造的

⑨ simulated /ˈsɪmjuleɪtɪd/　*adj.* 模拟的；模仿的

⑩ connectivity /kəˌnekˈtɪvətɪ/　*n.* 连通(性)；联结(度)

移动支付

Mobile Payment

导入语 Lead-in

移动支付是指使用移动设备进行付款的服务。在无需使用现金、支票或信用卡的情况下，消费者可使用移动设备支付各项服务或数字及实体商品的费用。移动支付将互联网、终端设备、金融机构有效地联合起来，形成了一个新型的支付体系，已经渗透到人们的吃穿住用行等各个方面。移动支付打破了传统支付对于时空的限制，使用户可以随时随地进行支付，并对个人账户进行查询、转账、缴费、充值等功能的管理。其便捷、高效的支付方式，成为越来越多消费者购物结账时的选择。

文化剪影　Cultural Outline

Mobile payment refers to payment services operated under the **financial**① regulations and performed by a mobile **device**②. Instead of paying with cash, cheque or credit cards, a consumer can use a mobile to pay for a wide range of services and digital or hard goods.

移动支付是指在金融监管下，通过移动设备进行的支付服务。与现金、支票或信用卡不同的是，消费者可以使用移动设备来支付各种服务和数字或实体商品。

Mobile payments have swept China. Whether shopping, buying a meal in a restaurant, taking a taxi, purchasing cinema tickets or paying **utility**③ bills, almost every financial transaction can be managed on a mobile phone by scanning a QR code.

移动支付已经席卷了中国。无论是购物、在餐馆吃饭、打车、买电影票还是支付水电费，几乎所有的金融交易都可以通过扫描二维码在手机上进行。

The popularity of mobile payment is the result of joint efforts in many fields such as telecommunications, finance, and security.

移动支付的普及是电信、金融、安全等诸多领域共同努力的结果。

佳句点睛 Punchlines

1. Mobile payment plays a crucial role in boosting the urban economy, improving employment and **accelerating**① consumption upgrading.

移动支付在刺激城市经济发展、促进就业、加快消费升级等方面发挥了关键作用。

2. The two e-wallet giants such as Alipay and WeChat Pay **dominate**⑤ China's domestic mobile payment market.

支付宝和微信支付这两家电子钱包巨头主导着中国国内移动支付市场。

3. China will step up efforts to promote mobile payments by smartphones, QR codes and bio-identification, meeting people's diverse needs in different consumption **scenarios**⑥.

中国将进一步加强智能手机、二维码、生物识别等移动支付的推广，满足不同消费场景下人们的多样化需求。

 情景对话 Situational Dialogue

A: Oops, I left my wallet in the dormitory.

B: Don't worry. Let me buy the subway tickets.

A: So ... that's OK?

B: Yes, this is mobile payment. If I may be honest, I can't remember the last time I carried cash.

A: Wow, it's really fast and convenient.

B: In China, mobile payment is described as one of the "New Four Inventions," along with high-speed railway, shared bicycles and online shopping. People are using it everywhere, from supermarkets to restaurants, from taxis to high-speed railways, and even at **snack**⑦ stands and vegetable markets.

A: Great! I'd like to have a try.

B: You can download an APP, Alipay or WeChat Pay, and connect your bank cards.

A: Then what I need to do is just scan and click, like you?

B: Yes. And what's more, the account details **log**⑧ all of your **transactions**⑨, so you can easily track how much you spent and what you bought.

A: 哎呀，我把钱包落在宿舍里了。

B: 别担心。让我来买地铁票吧。

A: 这就买好了吗？

B: 对呀，这叫移动支付。说实话，我已经不记得最近一次带现金是什么时候了。

A: 哇，这真是既快捷又方便。

B: 在中国，移动支付与高铁、共享单车和网上购物一起，被称为

"新四大发明"。从超市到餐馆,从出租车到高铁,甚至在小吃摊和菜市场,人们都在使用移动支付。

A: 太棒了!我也想试一试。

B: 你可以下载一款应用程序,支付宝或微信支付,并绑定你的银行卡。

A: 然后我需要做的就是像你一样扫描和点击确认?

B: 是啊。更重要的是,账户详细记录了所有交易,你可以很容易地查询你花了多少钱、买了什么东西。

生词注解 Notes

① financial /faɪˈnænʃl/ *adj.* 金融的;财务的

② device /dɪˈvaɪs/ *n.* 设备;装置

③ utility /juːˈtɪlətɪ/ *n.* 公共事务;实用

④ accelerate /əkˈseləreɪt/ *vt.* 使……加快;使……增速

⑤ dominate /ˈdɒmɪneɪt/ *vt.* 控制;支配

⑥ scenario /səˈnɑːrɪəʊ/ *n.* 场景;情节

⑦ snack /snæk/ *n.* 小吃;快餐

⑧ log /lɒg/ *vt.* 对……做记录;把……载入日志

⑨ transaction /trænˈzækʃn/ *n.* 交易;事务